Cognitive Semantics and Scientific Knowledge

Converging Evidence in Language and Communication Research

Linguists have taken a broad view of language and are borrowing methods and findings from other disciplines such as cognitive and computer sciences, neurology, biology, sociology, psychology, and anthropology. This development has enriched our knowledge of language and communication, but at the same time it has made it difficult for researchers in a particular field of language studies to be aware of how their findings might relate to those in other (sub-)disciplines.

CELCR seeks to address this problem by taking a cross-disciplinary approach to the study of language and communication. The books in the series focus on a specific linguistic topic and offer studies pertaining to this topic from different disciplinary angles, thus taking converging evidence in language and communication research as its basic methodology.

Works in the series are designed to appeal to linguists and non-linguists alike. Each work in the series is written and/or coordinated by an expert in a particular field.

Volume 4

Cognitive Semantics and Scientific Knowledge: Case studies in the cognitive science of science
by András Kertész

Cognitive Semantics and Scientific Knowledge

Case studies in the cognitive science of science

András Kertész
University of Debrecen

John Benjamins Publishing Company
Amsterdam / Philadelphia

The paper used in this publication meets the minimum requirements of American National Standard for Information Sciences – Permanence of Paper for Printed Library Materials, ANSI Z39.48-1984.

Library of Congress Cataloging-in-Publication Data

András Kertész
Cognitive Semantics and Scientific Knowledge : Case studies in the cognitive science of science / András Kertész.
p. cm. (Converging Evidence in Language and Communication Research, ISSN 1566–7774 ; v. 4)
Includes bibliographical references and indexes.
1. Semantics. 2. Cognitive grammar. 3. Science--Philosophy. 4. Knowledge, Theory of. I. Title. II. Series.

P325.K456 2004
401'.43-dc22 2004041196
ISBN 90 272 3890 1 (Eur.) / 1 58811 501 1 (US) (Hb; alk. paper)

John Benjamins Publishing Co. · P.O. Box 36224 · 1020 ME Amsterdam · The Netherlands
John Benjamins North America · P.O. Box 27519 · Philadelphia PA 19118-0519 · USA

Contents

Preface

For more than a decade I have been interested in the relationship between linguistics and the philosophy of science. After such a relatively long time of thinking about sophisticated problems and complicated quandaries one may justifiably feel the need to synthesize the result of one's efforts. Therefore, the present book was originally intended to summarize some of my findings which I think are instructive. However, as work on the manuscript proceeded, the original intention changed. I gradually realized that the manuscript would turn into a starting point for future considerations rather than present an overview of past achievements. Consequently, although the book recapitulates some of the ideas I put forward earlier, it should rather be considered as a novel and, in several respects, unconventional reflexion on the way linguistics and the philosophy of science may interact.

One of the most interesting aspects of current developments in linguistics is whether, and if so, to what extent, as a result of the cognitive turn old boundaries of research can be crossed. In particular, the book centers on the question of how far the scope of cognitive semantics can be extended towards capturing problems which traditionally the philosophy of science has dealt with, and where those limits beyond which it cannot transgress lie. The considerations which will yield an answer to this question neither aim at a legitimization of cognitive semantics, nor do they attempt a fierce refutation of the discipline; rather, they are concerned with a problem-oriented analysis of some of its prospects and limits.

Exploring the boundaries of a field of research is a risky enterprise. It involves experimentation, and experiments, as we know, may fail. Thus, the present book is deliberately intended to run all the risks which are inseparable from tackling the issues it focuses on.

My research was supported by the Hungarian National Research Fund (OTKA) grant no. T034662. My thanks are due to László Bene, Péter Csatár, Zoltán Kövecses, Gergely Pethő, Csilla Rákosi, Enikő Tóth and Zoltán Vecsey for critical comments on earlier versions of the manuscript. I would also like to thank two anonymous reviewers at John Benjamins for helpful suggestions.

I'm especially grateful to George Seel for improving my English. None of these people is responsible for the shortcomings of the present work.

Debrecen, October 2003 András Kertész

Introduction

1. Overview

The literature abounds in exaggerated reactions to the cognitive turn in linguistics. Some of these reactions go as far as to embrace with enthusiasm cognitive linguistics in general and cognitive semantics in particular, and to consider them revolutionary:

> The fundamental stance of cognitive linguistics may best be summarized in terms of two key issues: *the nature of linguistic meaning* and the nature of grammar. In the view of some cognitive scientists, the cognitive linguistics approach to these two issues constitutes a *revolution* in our understanding of how human language and cognition operate. (Tomasello 1999: 478; emphasis added)

At the other end of the scale one finds fierce rejections. For example:

> [...] the situation has arisen where many linguists begin calling themselves 'cognitive scientists'. This semantic mentalism is often complemented with a kind of 'reduction axiom': everything mental is physical, every event in the mind is (in fact) an event in the underlying brain etc. This seems to guard against the suspicion that what is going on is the old mentalism which has been seriously challenged by so many philosophers — the structures which are studied are ultimately tangible structures of the human brain. However, this is *illusory* — the structures postulated by linguists are clearly not results of studying the brain — the books which present them do not map neural synapses or anything of their kind (and, in fact [...], if they did so, *they would not be about semantics*). (Peregrin 1998: 255; emphasis by underlining in the original, italics added)

Given these antagonistic evaluations, the most important basic question that the working semantician has to ask is this:

(Q) What are the prospects and limits of cognitive semantics?

However, this problem — which the working semantician is constantly confronted with during his daily activity — is very difficult to grasp. Namely, (Q) may be called an *ultimate* question. Ultimate questions are well-known from the history of philosophy and the sciences: "What is the world?", "What are human beings?", "What is nature?" etc. The constitutive property of ultimate questions is that they have *no global solution*. This means that they cannot be solved without the addition of background assumptions which were not at our disposal at the time they were asked and which *narrow them down*. Such background assumptions restrict the number and scope of possible answers, for example by clarifing the terms to be used, by defining the data on the basis of which possible answers may be obtained, and by introducing particular methods. The solutions which are available in this way may be called *local*, because they are valid only with respect to the additional background assumptions mentioned; if these background assumptions are changed, the answers may change as well. The history of the sciences consists to a considerable extent in the localization of ultimate questions. For example, the ultimate question "What is nature?" is answered by specific biological, physical, chemical theories in that they relate the question to their particular local network of theoretical terms, empirical hypotheses, methods and data.

Accordingly, the aim of Part I of the present book is to introduce the basic background assumptions which motivate the narrowing down of (Q) to a more specific problem. The localization of (Q) is carried out along the lines of two considerations.

Firstly, since (Q) focuses on the boundaries of possibilities, it is plausible to restrict the scope of the problem to 'extreme' cases within *the subject matter* of cognitive semantics. *Scientific knowledge* is ideally suited for this purpose. On the one hand, because it is commonly considered as an extremely *clear paradigmatic example* of abstract concept formation. On the other hand, because

> [...] cognitive linguistics is explicitly concerned to use *language as a window on cognitive structures generally*, so that one can move freely and gradually from facts about *language* to facts about *human cognition* and further on to facts about *human life generally* [...] (Harder 1999: 196; emphasis added)[1]

In full accordance with this some of the most instructive approches to cognitive semantics *explicitly claim that the investigation of particular properties of scientific concept formation belongs, among other things, to their intended applications*. Thus we make the decision to narrow down *the subject matter* of the present study to some of those peculiarities of scientific knowledge which can be studied by using cognitive semantic methods. Nevertheless, this decision also means that we will inevitably have to touch on problems which traditionally the philosophy of science has been interested in. Thus, there is no avoiding the interdisciplinary interaction of cognitive semantics with the philosophy of science. As a result of this interaction the present work is to be regarded as a contribution to the new field of *the cognitive science of science*.

The *second* consideration is that *cognitive semantics* itself has to be narrowed down to those clear-cut cognitive semantic *theories* whose workability may be tested by applying them to the analysis of some interesting features of scientific knowledge. Therefore, two very different theories are chosen as plausible and instructive *paradigmatic examples*: namely, Lakoff and Johnson's cognitive theory of metaphor as an example of holism, and Bierwisch and Lang's two-level approach as an example of modularism.

As a result of these two considerations, we obtain the following localization of (Q): *What are the prospects and limits of the cognitive theory of metaphor and of the two-level approach to cognitive semantics, if they are applied to scientific knowledge?* This question is *the main problem* of the book and its solution is the main objective of the argumentation to be presented.

The most effective way of revealing the prospects and limits of scientific approaches is to *test them in the course of applications*. Therefore, Parts II–IV are based on a series of *case studies*.

Since both the cognitive theory of metaphor and the two-level approach claim to have made a significant contribution to lexical semantics, Part II asks the question of whether, and if so, to what extent they can capture one of the *classic* quandaries of the analytic philosophy of science which clearly concerns the semantics of the expressions scientific theories make use of: namely, the so-called 'problem of theoretical terms'. The answer will be that while the two approaches draw very different pictures of the nature of theoretical terms, at the same time they both reject the assumptions of the analytic philosophy of science.

After a discussion in Part II of a classic problem of the philosophy of science, Part III is devoted to an issue which is *currently* the subject of heated debates: in particular, the relationship between the conceptual and the social factors of scientific knowledge. The case studies show that the cognitive theory of metaphor and the two-level approach facilitate different sociological extensions and lead to different accounts of the way the social and the conceptual factors of scientific knowledge interact. The concluding chapter of Part III summarizes the *prospects* which the case studies in Parts II and III suggest and draws far-reaching conclusions.

The case studies in Part IV discuss some of the *limits* of cognitive semantics that become evident when it is applied to scientific knowledge: the problem of constructivity, the role of plausible inferences and the emergence of fallacies. In analogy to the way Part III was concluded, the last chapter of Part IV summarizes the limits that have been revealed throughout our reasoning.

The final chapter recapitulates the structure of the argumentation and puts forward the solution to the main problem of the book. The solution says, on the one hand, that the prospects of the two cognitive semantic approaches examined in this book *radically widen the scope of linguistic research*, because they seem to be capable of reformulating and solving classic problems and current quandaries in the philosophy of science. On the other hand, however, they cannot transgress certain *serious limits* which may call these prospects into question. The book concludes by pointing out how some of the central findings can serve as possible starting points for future research.

This line of argumentation has certain *unconventional* features, which are discussed in the next sections.

2. Interdisciplinarity

Cognitive science is, by definition, an interdisciplinary enterprise. In full accordance with this, the present work focuses on the interdisciplinary interaction of two fields of research, namely, cognitive semantics and the philosophy of science. *The book reveals in what way and to what extent cognitive semantics can serve as a means of solving some of the central problems of the philosophy of science.*

The complexity of the problems which this interdisciplinary interaction raises would ideally require two sets of conditions to be met if we are to deal with them satisfactorily. First of all, metatheoretical explicitness, valid inferences, the precise use of terminology, and empirical hypotheses. Secondly, the many-sided and extensive examination of both disciplines to uncover the depth of their problems, their sophisticated methods and the relevance of their findings. Unfortunately, however, neither of these desiderata can be satisfied at the outset. Both the philosophy of science and cognitive semantics are characterized by the pluralism of methods and aims, by the uncertainty of the empirical foundations, by alternative hypotheses, as well as by the coexistence of incompatible theories and the vagueness of basic notions. Moreover, it would be unrealistic to expect that the readers of this book are competent both in the philosophy of science and in cognitive linguistics to the same extent. The consequences of conducting interdisciplinary research with these restrictions affect the nature of the argumentation to be carried out at least in the following respects.

Firstly, the interdisciplinary nature of the book raises the question of who the intended readers are. Primarily, it addresses *linguists*, who may be interested in *how far the scope of their discipline can be extended and where the limits which it cannot transcend lie.* Nevertheless, besides linguists, philosophers of science may also enrich their tools of analysis by the considerations which cognitive semantics as a possible approach to the new field of the cognitive science of science offers them.

Secondly, the argumentation sketched in the previous section showed that (Q) would be reduced to examining how two cognitive semantic approaches can be applied to the analysis of the structure of scientific knowledge. This touches on the question of *what exactly the book is about*: Is it about cognitive semantics or scientific knowledge? The answer is, of course: *about both*. As we all know, the analysis of a certain subject matter — in our case scientific knowledge — by a certain tool — in our case cognitive semantic theories — reveals both the properties of the subject matter and the effectivity of the tool. Therefore, discussing certain features of scientific knowledge by the use of cognitive semantics naturally yields an insight into the prospects and limits of the latter, too.

Thirdly, the interdisciplinarity of the argumentation to be developed here leads, among other things, to an essential difficulty. At relevant points of our reasoning we should presuppose background knowledge of the particular field at issue — that is, of several, very different trends within the philosophy of science, within cognitive linguistics in general and within cognitive semantics in particular. Therefore, it would be necessary to introduce this background knowledge by a *state-of-the-art analysis* in order to provide conclusions serving as the premises of our later argumentation. However, we will decide on another approach: we will summarize the background knowledge presupposed in the form of *clearcut sets of theses* which will be documented by the use of references to and quotations from well-known textbooks and frequently cited seminal works. Hopefully, the relatively simple presentation of highly complex issues will not lead to a simplistic picture of particular theories, trends, and views, and will not undermine the credibility of our main findings, either. There are two reasons for choosing this strategy. The first is that detailed state-of-the-art analyses concerning all those fields of research whose interaction the present book focuses on would go far beyond the scope of the study itself. Secondly, some of the case studies, because of the technical methods to be applied and the nature of the problems to be tackled, will involve complicated and sophisticated argumentation; therefore, we would be well advised to keep at least the points of departure *as simple and as comprehensible as possible*.

Fourthly, there is no avoiding the fact that precise definitions of central notions will not, as a rule, be possible. Basically, all notions will be used deliberately in a *pre-explicative* sense. Nevertheless, in certain cases it will still be necessary to clarify the terminology to be applied. Any attempts at clarification of terminology will be motivated by the given context of argumentation and are not claimed to lead to generally valid definitions or explications. Even this very cautious treatment of the terminology will hopefully be sufficient to support important conclusions and substantial theses.[2]

Fifthly, the reader is expected to reflect carefully on the structure of the argumentation. The argumentation proceeds linearly and hierarchically, because earlier *conclusions* are used later as *premises* of further inferences. Therefore, the premises, the key notions and the structure of the inferences always have to be as clear as possible. This means that, for the sake of clarity, theses, premises, terms, and conclusions already introduced will be referred to at later stages of the argumentation.

3. The case studies as thought experiments

Another important peculiarity of the book is that the investigations summarized in Section 1 above will be reduced to a series of case studies, whose task is to illustrate the issues which may lead, in the final analysis, to a possible solution to the main problem. Constructing such case studies brings with it at least two fundamental difficulties. The first is that since cognitive semantic approaches consider themselves to be empirical, in principle the case studies should play the role of empirical experiments. Unfortunately, however, as is well-known, the empirical basis of most cognitive semantic theories is at present modest and uncertain. Secondly, scientific knowledge is something highly complex and, despite the efforts of the philosophy of science over the past century, we know very little of how it is structured and operates. From these two difficulties — that is from the complexity of the subject matter and the weak empirical foundation of the methods of the investigation — it follows that the case studies which are expected to test the applicability of cognitive semantic theories to scientific knowledge have to be based on *thought experiments* rather than on real experiments. It is important to emphasize that this is not a deficiency of the present book, but *a constitutive property of the state of the art.* Therefore, the prospects and limits of the metascientific application of cognitive semantics will be closely related to the way thought experiments can be carried out.

Although thought experiments are, in certain contexts, very fruitful tools of scientific and philosophical problem solving, there is no agreement about what they are. See, for example, Peijnenburg and Atkinson's concise summary of the state of the art:

> [...] a lively debate on the nature of thought experiments can be discerned in the literature from 1990. Thought experiments have been defined as limiting cases of experiments (Sorensen 1922), as arguments (Norton 1991, 1996), as "guided contemplations" (Gendler 1998), as vistas in a platonic world (Brown 1991), as specific functions in an experiment (Borsboom et al. 2002), or as not arguments at all (Bishop 1999). However, this disagreement about what thought experiments *are* contrasts with the unanimity about what thought experiments should *do*; and it is the latter, not the former, that counts if we want to distinguish good from bad thought experiments.
>
> In the view of almost everyone, a thought experiment should give sudden and exhilirating insight. (Peijnenburg and Atkinson 2003: 305–306)

Instead of going into a discussion of the literature, we will take Cooper's (1999) considerations about the nature of thought experiments in the philosophy of mind as a point of departure. With respect to the case studies to be carried out in this book, the following features of thought experiments are worth mentioning:

Firstly,

> a thought experimenter gains knowledge through *manipulating a model* (Cooper 1999: 267; emphasis added).

Secondly,

> in a thought experiment we manipulate models in accordance with the *answers to 'What if?' questions.* [...] When all the 'What if?' questions are followed through, the result of the manipulations is either an internally consistent model, constituting a possible world, or contradiction. If the thought experimenter achieves an internally consistent model, then he can conclude that the situation he has imagined is *possible.* If the thought experiment ends in a contradiction then the situation is *impossible.* (Cooper 1999: 268; emphasis added)

Thirdly, thought experiments use not only deductive, but also *non-deductive reasoning.*[3] Fourthly,

> the form of the model is *unconstrained.* The model may be a mental model in someone's head, or it may be a computer simulation, or it may be made from plasticine and <u>papier mâché</u>. [...] Modelled phenomena do not necessarily unfold as they would in the real world, as the thought experimenter may model a world in which some laws of nature are *suspended* or *altered.* (Cooper 1999: 267–268; emphasis by underlining in the original, italics added).[4]

Accordingly, the two-level approach and the cognitive theory of metaphor will serve as *models to be manipulated in the course of their application to scientific knowledge.* Each case study will ask the question of what the case would be *if* the two-level approach or the cognitive theory of metaphor were accepted. In answering these questions, there will be *no prefabricated constraints on the methods of argumentation*, that is, besides the use of deductive inferences, the application of the rich inventory of non-deductive *reasoning* is also permitted. Thereby, the information we are allowed to use is in principle *unconstrained*, too, in that we will be free to utilize any background assumptions which at certain points of the argumentation seem to be motivated by *heuristic* considerations. Our conclusions as to the prospects and limits of the two approaches mentioned will be drawn from examining in what situations and to what extent the manipulation of the models will lead to serious *difficulties* such as contradictions or other kinds of limits.

It follows from what has been said that the task of the present author is twofold: he is the thought experimenter, and at the same time he plays the role of the cognitive semantician. Therefore, his attitude is that of *self-reflexion.* On the one hand, the subject matter of the case studies is scientific knowledge. On the other hand, the way certain aspects of scientific knowledge are investigated in the present work *is itself* an example of how scientific inquiry proceeds under conditions characterized, for instance, by the insufficiency of background information, by incomplete data, by uncertain premises and the non-deductivity of argumentation.

It is also useful to remark that all the case studies are constructed in such a way that they reveal *both* some of the prospects and some of the limits of the approaches to be tested. Nevertheless, so as to avoid the overcomplication of the argumentation, in the case studies in Parts II and III only the prospects will be evaluated, whereas the limits will be summarized in Chapter 14.

4. Summary

The present book has the following special characteristics:

- It attempts to reveal some of the prospects and limits of cognitive semantic theories by applying two of them to scientific knowledge.
- It integrates information from the philosophy of science and cognitive semantics in a strongly interdisciplinary manner.

– The main problem of the book will be solved by carrying out case studies which are to be regarded as thought experiments.
– Due to the nature of thought experiments, the argumentation will proceed self-reflexively.

As a result of these basic features of the present work, at every point of the argumentation we will have to make decisions which may be uncertain or may be seriously questioned. Therefore, the very nature of this enterprise makes it appear hazardous at the outset. Nevertheless, the perspective that through our efforts we may learn something about the highly intricate mechanisms governing scientific inquiry undobtedly makes it worth running *all the risks* which are associated with the considerations put forward in the following pages.

Finally, let us add a few technical remarks. Throughout the book certain *conventions* have to be made use of. Italics indicate emphasis and elements of object-language are italicized as well. Expressions, whose occurrence is relevant in a given context but which are used in a pre-explicative sense, are (occasionally, but not mechanically in each case) put within single quotation marks. Important problems and hypotheses will be labelled by abbreviations — nevertheless, in referring to them their full names are also used. The theses which are often referred to will be listed in the Appendix so that the reader can identify them more easily.

Part I

Preliminaries

Chapter 1

On the cognitive turn

1.1 Introduction

In this chapter our main task is to arrive at presentations of basic notions which are *simple and clear* enough to serve as the background for the relatively complicated later analyses. Section 1.2 will be devoted to the cognitive turn in linguistics. In Section 1.2.1 we will start with a heuristic characterization of *cognitive* science. On the basis of this, in Section 1.2.2 one possible account of the notion of cognitive linguistics will be outlined. We will restrict our attention to the relationship of two main trends within the latter, namely, *modularism and holism* (Section 1.2.3). In Section 1.2.4 these two trends will be specified to cognitive semantics. Although they do not characterize cognitive semantics exclusively, they are still dominant enough to yield instructive case studies which will help us to test some of the prospects and limits of the latter. In Section 1.2.5 (Q) will be narrowed down to another problem (Q′) which specifies cognitive semantics to the relationship between modularism and holism.

In the *Introduction* we said that the prospects and limits of cognitive semantics would be discussed by considering case studies which focus on the cognitive aspects of *scientific* inquiry. Since it is the philosophy of science which deals with the nature of scientific inquiry, as a next step we have to *establish the connection* between this possible application of cognitive semantics and current trends in the philosophy of science. That is, in Section 1.3 we have to consider the question of whether, and if so in what way, current trends in the philosophy of science legitimatize the use of cognitive semantic methods to capture certain aspects of scientific knowledge. We will see that it is the *naturalized philosophy of science* that is expected to yield an explicit answer to this question. After a brief terminological digression in Section 1.3.1, in Section 1.3.2 we will focus on the state of the art in the naturalized philosophy of science. Section 1.3.3 will be devoted to one particular manifestation of the latter, namely, *the cognitive science of science.* Section 1.3.4 will draw those conclusions which suggest the reformulation of (Q′) as the more specific problem (Q″) focussing on the prospects and limits of the metascientific extension of holistic and modular cognitive semantics.

Finally, in Section 1.4 the upshot of these considerations will be summarized and at the same time the next step of argumentation prepared.

1.2 On the cognitive turn in linguistics

1.2.1 Cognitive science

According to Howard Gardner's classic historiographical overview of the first decades of cognitive science, the nature of the cognitive turn can be characterized in a pointed and concise way. Firstly, one of the main tasks of cognitive science is to reformulate, to answer or to eliminate classic philosophical questions concerning the nature of the mind and knowledge by using the methods of empirical disciplines.[1] It is important to realize that in Gardner's view it is this transformation of classic philosophical questions into empirical ones and the attempt to solve them empirically which, apart from other important aims, is *central* to the notion of cognitive science.[2] Secondly, cognitive science involves the cooperation of several subdisciplines, and *linguistics* is one of these subdisciplines.[3]

So as to demonstrate that since the publication of Gardner's classic work these basic assumptions of cognitive science have not been superseded by later developments over the past almost two decades, but are still very much valid, a passage from a more recent overview may be quoted:[4]

> However vague the theoretical foundations of this new science are, there is still agreement on its year of birth: 1956. At that time Simon, *Chomsky*, Newell and others met at MIT at the Symposium on Information Theory and started a new way of inquiry into cognition. The novelty consisted in the attempt *to transform very old philosophical questions* concerning the nature of the mind into the question of how it works and to answer the latter interdisciplinarily and empirically. As uncontroversial as its year of birth is the fact that the following disciplines belong to cognitive science: *linguistics*, computer science, neurobiology, and the philosophy of mind. (d'Avis 1998: 37; emphasis by underlining in the original; my translation, A. K.)

With respect to linguistics, the quotation illustrates the fact that cognitive science has, among other things, two very important properties:

(1) a. One of the central aims of all subdisciplines of cognitive science is to reformulate, to answer or to eliminate classic philosophical questions

concerning the nature of the mind and knowledge by dealing with them empirically and interdisciplinarily.

b. Linguistics is undoubtedly one subdiscipline of cognitive science which is expected to proceed in a basically interdisciplinary manner.

(1a) and (b) motivate the following inference:

(2) *If*

a. a certain kind of linguistics is one of the subdisciplines of cognitive science, and

b. one of the central aims of all subdisciplines of cognitive science is to reformulate, to answer or to eliminate classic philosophical questions concerning the nature of the mind and knowledge by dealing with them empirically and interdisciplinarily,

then

c. one of the central aims of a certain kind of linguistics is to reformulate, to answer or to eliminate classic philosophical questions concerning the nature of the mind and knowledge by dealing with them empirically and interdisciplinarily.

Now, in accordance with (2), the following tentative picture of *the linguistic subdiscipline of cognitive science* suggests itself:

(3) The linguistic subdiscipline of cognitive science is such that

a. its *object* of investigation is knowledge of language as part of cognition,

b. its *methods* are those of empirical linguistics, and

c. its *aim* is, among other things, to reformulate, to answer or to eliminate classic philosophical questions concerning the nature of the mind and knowledge by dealing with them empirically and interdisciplinarily.

On the basis of what has just been said we may consider (3) to be the common background assumption of all linguistic approaches whose objective is to contribute to cognitive science. Therefore, (3) applies, among other things, to approaches which may be antagonistic in many respects but which share some basic tenets of cognitive science. It is instructive to illustrate this fact by two quotations which represent two very different paradigmatic examples within the linguistic subdiscipline of cognitive science. With respect to generative linguistics, Chomsky expresses exactly the ideas put forward in (3):[5]

> One might ask whether there really is a '*scientific study of language*'. My own view is that such a field is beginning to take shape. [...] On the basis of these glimmerings of understanding, it seems that there may well be significant implications with

> respect to certain *classical problems of philosophy.* My own guess is that these implications may prove to be the richest with respect to the theory of *human knowledge* and understanding and more generally with respect to the nature of *mind.* In particular, what we are now coming to understand suggests that some of the questions of the theory of knowledge should be recast.[...]
>
> In my opinion, it is within this larger framework that the technical developments that have taken place within the field of *generative grammar* should be understood. (Chomsky 1986: 271–3; emphases added)

Although Lakoff and Johnson reject the hypotheses and the methods of generative grammar, in their latest book they outline their enterprise in a way which is also in full accordance with (3).

> In Part II, we study *the cognitive science of basic philosophical ideas.* That is, *we use these methods* to analyze certain basic concepts that any approach to philosophy must address [...].
>
> In Part III, we begin the study of philosophy itself from the perspective of cognitive science. We apply these analytic methods to *important moments in the history of philosophy.* [...] *These methods, we argue, lead to new and deep insights into these great intellectual edifices.* [...] We also take up issues in *contemporary* philosophy [...].
>
> What emerges is a philosophy close to the bone. A philosophical perspective based on our *empirical* understanding of the embodiment of mind is *a philosophy in the flesh,* a philosophy that takes account of what we most basically are and can be. (Lakoff and Johnson 1999: 8; emphases added)[6]

Consequently, after having characterized the linguistic subdiscipline of cognitive science, within this we have to distinguish between 'cognitive linguistics' and generative grammar. In other words, with respect to 'cognitive linguistics', (3) outlines the *genus proximum*; now we will try to find out the *differentia specifica.*[7] The next section will focus on the possibility of such a distinction by introducing the term 'cognitive linguistics'.

1.2.2 Cognitive linguistics

For some commentators, the beginnings and the importance of cognitive linguistics appear to be as clear as those of cognitive science (according to d'Avis 1998). The identification of 'cognitive linguistics' may simply rest on a very general basic tenet concerning *the cognitive aspects of communication*, the publication of a set of seminal works and the birth of institutions:[8]

> [...] *cognitive linguistics* refers to the set of theories that are primarily concerned with the cognitive dimensions of linguistic communication. Although there were

> important precursors in the work of linguists such as Charles Fillmore and Leonard Talmy, cognitive linguistics had its clear origins as a scientific paradigm in 1987 with the publication of George Lakoff's *Women, Fire and Dangerous Things: What Categories Reveal about the Mind* and the first volume of Ronald Langacker's *Foundations of Cognitive Grammar* — followed immediately by the founding of the International Cognitive Linguistics Association and its official journal *Cognitive Linguistics*. (Tomasello 1999: 477–478; emphases in the original)

Some of its additional characteristics are as follows:

> The most general point [...] is that *it is basically impossible to isolate linguistic meaning from cognition in general* in the manner of a mental lexicon divorced from other aspects of human cognition and communication. Cognitive linguistics therefore adopts an encyclopedic, *subjectivist* approach to linguistic meaning in which human beings create and use linguistic conventions in order to symbolize their shared experiences in various ways for specific communicative purposes. These different experiences and purposes are always changing: so they can never be captured by an itemized, *objectivist* description of linguistic elements and their associated truth conditions. For an *adequate description of linguistic semantics from the cognitive linguistics point of view*, what is needed is a psychology of language in terms of such things as cognitive structures, the manipulation of attention, alternative construals of situations, and changing communicative goals. [...] This view of linguistic communication and the cognitive processes on which it depends is obviously *very different from that of generative grammar* and other formalistic approaches. (Tommasello 1999: 480–484; emphasis added)

An important observation is that 'cognitive linguistics' appears, in the light of these quotations, to be a network of theories which is clearly distinguished from the generative tradition, although both belong to the linguistic subdiscipline of cognitive science.[9] Thus, 'cognitive linguistics' is conceived of *in opposition to* generative linguistics (within the linguistic subdiscipline of cognitive science). The main point is that while the latter maintains the *autonomy* of the knowledge of language in general and that of syntax in particular, the central feature of cognitive linguistics is seen to be the denial of this kind of autonomy. Accordingly, we may conclude (4):

(4) Cognitive linguistics as a subdiscipline of cognitive science
  a. accepts (3a–c), and
  b. comprises a set of theories that assume that knowledge of language and other cognitive processes constitute a *unified* system.

However, there is *another view* of cognitive linguistics as well. In her classic introduction, M. Schwarz characterizes the state of the art as follows.[10]

> By way of a rude differentiation, within *Cognitive Linguistics* we may distinguish between two trends, which reveal a kind of dualism determining the whole of cognitive science research: *the modular and the holistic approach.* Not all works which have recently been put forward within Cognitive Linguistics can be mapped onto one of these two positions.[...] *But basically Cognitive Linguistics is literally divided into two 'camps'.* The rigid exclusiveness with which one or another view is maintained certainly handicaps their future cooperation. (Schwarz 1992: 44; my translation, A. K.; emphasis added)

We will accept this terminology; to refer to the dichotomy mentioned in the above quotation, we will use the terms 'modularism' and 'holism'. However, this choice of terminology is not vitally important. We could have used Langacker's distinction between 'formalism' and 'functionalism', or Lakoff and Johnson's terms 'first generation' versus 'second generation cognitive science' — although these dichotomies are not strictly equivalent but highlight slightly different aspects. The main reason for our terminological decision is philological and historiographical rather than substantial. As Müller's (1991) and Gardner's (1985) historiographical studies convincingly showed, the dichotomy between these two "camps" is rooted very deeply in the history of the sciences in general and of what today is called 'cognitive science' in particular. In these studies the terms 'holism' and 'modularism' are used. The other dichotomies mentioned are applied by the proponents of a certain theory within cognitive linguistics so as to distinguish their particular approach from views which they reject; accordingly, they are strongly theory-dependent and serve the purposes of particular approaches. Therefore, it is plausible to choose a more neutral terminology motivated by a wider historiographical and philological context.

Now, what Schwarz calls the 'holistic' approach clearly *corresponds to (4)*, and what she labels the 'modular' approach includes theories which are *compatible with* generative grammar in that they, just like the latter, accept the autonomy of the knowledge of language. One of the paradigmatic examples of modular cognitive linguistics is the so-called 'two-level semantics' as represented e.g. in Bierwisch (1983a), (1983b), (1987), (1990), Bierwisch and Lang (eds.)(1989), Bierwisch and Schreuder (1992), Lang et al. (1991); another is Jackendoff's 'conceptual semantics' (see e.g. Jackendoff 1983, 1990 etc). Paradigmatic examples of holism are all those publications which were listed in note 8 and in the first quotation by Tomasello in this subsection.

Thus in this second sense we may characterize the notion 'cognitive linguistics' as follows.

(5) Cognitive linguistics as a subdiscipline of cognitive science
  a. accepts (3a–c), and
  b. comprises at least two sets of theories, namely, holistic and modular approaches.

(4) and (5) cannot be maintained simultaneously, thus we have to choose between them.[11] Since our task in this book is to explore the prospects and limits of cognitive semantics (as indicated in (Q)) by analyzing instructive cases (see the Introduction), the second interpretation of the term mentioned in (5) seems to be much more useful, because it allows us to contrast two antagonistic approaches to the knowledge of language. Therefore, we suggest the following *convention*:[12]

(6) The term 'cognitive linguistics' will be used in the sense of (5).

In the light of (5) and (6), as a next step we have to consider the two main trends within cognitive linguistics, namely, holism and modularism.

1.2.3 Modularism vs. holism

Let us start from one of the basic questions of linguistics:

(7) What is the knowledge of language?

(7) is an ultimate question in the sense mentioned in the *Introduction*. Therefore, we have to assume that, since it has no *global* solution, the task of the linguist is to determine the specific conditions on the basis of which *local* solutions may be obtained. Such a background assumption aiming at the localization of the problem is the central empirical hypothesis of cognitive linguistics already mentioned in (3)(a):

(8) Knowledge of language is part of cognition.

So, we get the following localization of the ultimate question (7):

(9) What is knowledge of language, *if* knowledge of language is part of cognition?

Basically, though not exclusively, cognitive linguistics has given two answers to this question which seem to influence current research in a constitutive way: the holistic and the modularistic answer. Let us examine the holistic answer first.

(10) *The holistic answer to (9):* Knowledge of language and other cognitive processes constitute a unified (holistic) system.

The following quotations may illustrate the idea reconstructed in (10).

> Grammar (or syntax) does not constitute an autonomous formal level of representation. […] There is no meaningful distinction between grammar and lexicon. Lexicon, morphology, and syntax form a *continuum* of symbolic structures, which differ along various parameters but *can be divided into separate components only arbitrarily.* (Langacker 1987: 2–3; emphasis added)

> The distinction between semantics and pragmatics (or *between linguistic and extralinguistic knowledge*) is largely artefactual, and the only viable conception of linguistic semantics is one that avoids such false dichotomies […]. (Langacker 1987: 154; emphasis added)

> It [Lakoff and Johnson's approach, A. K.] includes, for starters, *all those things* you would have to learn if you were to learn a foreign language: the meanings, the pragmatics, the speech-act constructions, constraints on processing, and on and on. (Lakoff and Johnson 1999: 482)

The holistic hypothesis formulated in (10) may be further subdivided into the following claims:[13]

(10) a. *The thesis of complexity.* According to (10), cognition does not consist of autonomous subsystems. Accordingly, knowledge of language is determined by human intelligence in general rather than by a language faculty assumed to be independent from other mental faculties. Moreover, it is not possible to distinguish between knowledge of language and knowledge of the world. Finally, knowledge of language thus characterized is itself indivisible and does not consist of separate, autonomous subsystems.

b. *The thesis of nonformalizability.* Our scientific assumptions concerning knowledge of language as part of cognition (see (8)) cannot be formalized.

c. *The thesis of prototypicality.* Accordingly, semantics does not assume the existence of discrete set-theoretic entities, but instead rests on categorization along the lines of prototypicality.

d. *The thesis of the irrelevance of truth conditions.* The meaning of complex expressions cannot be captured on the basis of truth conditions.

e. *The thesis of conceptualization.* There is no dividing line between conceptual and semantic structures. Semantic structures are determined by the perspective of conceptualization.

f. *The thesis of encyclopedic systematization.* The traditional lexicographic systematization of word meaning must be replaced by encyclopedic systematization.

g. *The thesis of cognitive principles.* Semantic structures must be described with respect to cognitive domains. Grammar is governed by cognitive domains.

h. *The thesis of metaphorization.* Metaphorization is relevant for the organization of knowledge.

The modularistic answer to (9) says the opposite of (10):

(11) *The modularistic answer to (9)*: Knowledge of language is a relatively autonomous system (=module) interacting with other systems and is itself structured by the interaction of relatively autonomous (sub)systems.[14]

One of the paradigmatic examples of modular cognitive linguistics is the two-level semantics elaborated by M. Bierwisch and E. Lang which presupposes the basic tenets of generative linguistics and which supplements the syntactic component of the latter, among others, with a semantic and a conceptual one. The two-level semantics puts forward the modularity hypothesis in this way:[15]

> Basically, *all human cognitive behaviour* is organized in a modular fashion. The structure formation underlying any concrete behaviour performance is *based upon* the integration of various relatively autonomous, task-specifically interacting systems and subsystems (MODULES). *Language*, the different modes of perception, and the conceptual organisation of experience make up such systems, *which for their part are again structured in a modular way.* The *aim* to be derived from this assumption is to identify systems, to analyze their structure and organisation in the attempt to capture the rationale behind their interaction. (Lang and Carstensen, 1990: 6; emphasis added)

By analogy to (10), (11) may be subdivided into the following more specific assumptions which are, however, the opposites of the corresponding tenets in (10a–h).

(11) a. *The thesis of simplification.* For methodological reasons the object of investigation — as a consequence of its complexity — can be captured only via systematic simplifications. The most effective way of simplifying the treatment of the object of investigation is by subdividing it into subsystems. If we try to subdivide cognition into subsystems, hypothetically we may speak for example of the system of visual perception, the system of auditory perception, the motoric

system, the conceptual system etc. One of these systems is knowledge of language which may be further broken down into several subsystems — such as, for example, the systems of phonological representations, morphosyntactic representations and semantic representations.

b. *The thesis of relative autonomy.* All these systems are *autonomous* in so far as they are not reducible to other systems. At the same time, however, this autonomy is *relative*, because the systems interact with each other via interfaces in a well defined way. Such relatively autonomous systems are called *modules.*
c. *The thesis of formalization.* We have to formalize our scientific assumptions concerning the nature of these systems.
d. *The primacy of theories.* The empirical data have to be interpreted by a theoretical framework which is presupposed at the outset.
e. *The thesis of set-theoretic structures.* Semantics presupposes the existence of discrete entities, i.e. set-theoretic structures.
f. *The thesis of truth conditions.* Semantics is based on truth conditions.
g. *The thesis of the autonomy of conceptualization.* Semantic and conceptual structures belong to different modules.
h. *The thesis of the two levels of word meaning.* Therefore, the meaning of lexical units results from the interaction of two modules: meaning is the context-dependent mapping of semantic representations onto conceptual structures.
i. *The secondary nature of metaphorization.* Metaphorical structures are of minor importance for the organization of knowledge.

1.2.4 Cognitive semantics

Having surveyed the two main trends within cognitive linguistics, a further term has to be clarified, namely, 'cognitive semantics', which (Q) focuses on.[16] As Tomasello (1999) emphasized in the second passage quoted in Section 1.2.2 and the quotation in the Introduction, one major subfield of cognitive linguistics is the study of 'meaning' i.e. 'semantics'. In accordance with (3), (5) and (6), by cognitive semantics we will mean:[17]

(12) *Cognitive semantics* is that subfield of cognitive linguistics

a. whose object of investigation is the 'meaning' of linguistic expressions as part of cognition;
b. whose methods are those of empirical linguistics;
c. whose central aim, among others, is to reformulate, to answer or to eliminate classic philosophical questions concerning the nature of

the mind and knowledge by dealing with them empirically and interdisciplinarily; and
d. which includes both modular and holistic approaches.

1.2.5 Conclusions

Given the current state of the art, it is only natural that the problem (Q) we raised in the Introduction cannot be captured in a unified and general way. Therefore, we have to reduce it to a *more specific* question. Let us replace, therefore, (Q) by (Q′), where the latter differs from the former in that it replaces the vague term 'cognitive semantics' by an interpretation which consists of at least two trends, namely, holism and modularism:

(Q′) What are the prospects and limits of modular and holistic cognitive semantics?

Though at this point the subject matter of the investigations to be carried out seems to be sufficiently narrow, (Q′) is still not specific enough to be tackled successfully. In the *Introduction* we said that the prospects and limits of cognitive semantics will be tested by examining in what way and to what extent the methods of cognitive semantics are capable of revealing relevant aspects of *scientific knowledge*. Therefore, what we need as the next step is a short overview of the state of the art of the discipline which is primarily concerned with the investigation of scientific knowledge, namely, the philosophy of science.

1.3 On the cognitive turn in the philosophy of science

1.3.1 Terminological preliminaries

As mentioned in Section 1.1, it is the naturalized philosophy of science that facilitates the connection between the possible application of cognitive semantics to the analysis of some properties of scientific knowledge and the philosophy of science itself. Before introducing the notion of naturalized philosophy of science, for the sake of clarity, it is necessary to make a short terminological digression.

Firstly, since the present study focuses on the interaction between cognitive semantics and the philosophy of science, after we have clarified what is generally meant by the former, it would be in order to present a simple characterization of the latter as well so as to provide a suitable point of departure. However, this is by no means an easy task:

> The prospects of giving an enlightening characterization of the philosophy of science in terms of its constituent notions are *bleak*. It will be more efficient to consider *what those who call themselves philosophers of science actually care about and do*. There is no hiding the fact that they are an eclectic lot who *do a diverse range of things*, some of them strange. (Newton-Smith 2001a: 3; emphasis added).

In spite of this, it seems to be possible to elucidate what philosophy of science is if we try to enumerate this "diverse range of things". For example:

> In addition to considering the aims, methods, and tools of science, *the philosopher of science is interested in the products of science, the contents of its theories* — in particular, those very general, powerful theories which offer pictures of the world radically at odds with our commonsense view of things. (Newton-Smith 2001a: 5–6)[18]

Secondly, in accordance with what has just been said, it is necessary to touch on the difference between the notions 'epistemology' and 'philosophy of science', even if a comprehensive explanation is impossible. A brief characterization is all the more important, because it is not only the philosophy of science, but also epistemology which is interested in the nature of scientific inquiry. Although its scope has been much wider, in accordance with current views which are closely related to our later argumentation, we will narrow down the use of the term 'epistemology' to the 'foundations of science'.[19] While the perspective of epistemology focuses on the "foundations" of scientific knowledge — i.e. questions like 'What are the conditions of scientific knowledge?', 'How does scientific knowledge come into being?', 'How is scientific knowledge related to the world'? etc. —, the subject matter of the philosophy of science is the structure and content of existing scientific theories rather than such general foundational problems of inquiry. We will focus on the philosophy of science in this sense and *concentrate on some relevant properties of scientific knowledge as manifested in scientific theories.*

Thirdly, we will presuppose the distinction between 'objectscience' and 'metascience' and occasionally use the term 'metascience' (and 'metascientific') as a shorthand for 'philosophy of science'.[20]

1.3.2 The naturalized philosophy of science

After this terminological digression let us start by noting that it was the analytic philosophy of science that dominated the scene for many decades and that Quine's naturalism was proposed as a reaction to the problems which the analytic philosophy of science could not account for. Therefore, naturalism is best understood by being contrasted with the analytic philosophy of science.

Although the analytic philosophy of science can be divided into several schools and although it has undergone considerable changes, there are certain basic assumptions which all the different schools and all the different phases of development share. In a very simplified manner we may summarize some of the central theses of the 'received view' (see e.g. Putnam 1962, Suppe 1970) of the analytic philosophy of science as follows.

(13) a. The analytic philosophy of science is a *philosophical* discipline.
b. The way it reflects on scientific inquiry presupposes certain *a priori* assumptions concerning the principles of rationality.
c. It strives to *justify* the results of scientific inquiry in so far as on the basis of the principles of rationality it evaluates scientific knowledge as true or false, rational or irrational, justified or not.

However, despite dominating the field for several decades, the decline of the analytic philosophy of science was inevitable. Among other things, it was Quine's *replacement thesis* which outlined a new perspective (Quine 1969a, Kornblith 1985). The starting point for this new perspective in the philosophy of science was Quine's proposal that traditional epistemology should be replaced by empirical science which, instead of justifying its object of investigation, should rather describe and explain it; thus epistemology should be 'naturalized'.[21] Before proceeding, we need to clarify the relationship between Quine's program of naturalized *epistemology* built on this claim and its consequences for the naturalized *philosophy of science.*

The relationship between epistemology and the philosophy of science is especially complicated with respect to naturalism as initiated by Quine's seminal *Epistemology Naturalised* (Quine 1969a). Although Quine was concerned with epistemology, during the past three decades his ideas were adapted by the philosophy of science as well — with modifications which are, among other things, closely related to the differences between the perspective of epistemology and that of the philosophy of science mentioned above. It is important to emphasize that, although Quine speaks of 'epistemology naturalized', in the literature the features which distinguish naturalized epistemology from traditional epistemology are assumed to be basically the same as those which distinguish traditional (analytical) philosophy of science from the 'naturalized philosophy of science' as well.[22] Thus the differences between 'epistemology' and the 'philosophy of science' are retained with respect to naturalism, although in a more specific sense. This means that whenever we will speak of 'naturalized philosophy of science', we will presuppose that the latter is *epistemologically motivated.* This means that the naturalized philosophy of science intends to adapt problems and solutions to these problems with respect to the specific aims and subject matter of the philosophy of science which were originally raised and suggested by Quine with respect to the perspective of epistemology. Consequently, it will be justified to motivate some of our assumptions which are within the scope of the philosophy of science with Quine's views rooted in epistemology. This strategy will rest on *analogical inferences.*[23]

The direct consequence of Quine's main thesis is the *reflexivity* of scientific inquiry according to which (not only epistemology, but also) the philosophy of science "is only *science self-applied.*" (Quine 1969b: 293; emphasis added). This new attitude resulted in what has been called the *naturalized philosophy of science.*

(14) The main tenets of the naturalized philosophy of science are these:[24]
  a. The philosophy of science must *not* be a philosophical discipline; rather, it should apply *scientific* methods.
  b. Accordingly, the philosophy of science should proceed in an *empirical, a posteriori* manner.
  c. Thus the task of the philosophy of science is to *describe and to explain* the results of scientific inquiry by using the a posteriori/ empirical methods of the sciences in accordance with (a) and (b).

As a first approximation, the relationship between the analytic and the naturalized philosophy of science can be stated in the following way.

(15) a. By using the a posteriori, empirical methods of '*science*' itself, the naturalized philosophy of science claims to solve or to eliminate the problems which the analytic philosophy of science tackled on the basis of its justificatory methodology.
  b. Accordingly, the analytic philosophy of science should be '*replaced*' by science itself.

Nevertheless, so as to avoid misrepresenting the situation, this picture should be subjected to an at least twofold refinement. On the one hand, we should ask the question: What is meant by '*science*' in (15a)? That is, we have to find out what it is that is expected to replace the analytic philosophy of science. On the other hand, it is anything but clear what we mean when we say that 'science', whatever it may be, should '*replace*' the analytic philosophy of science as formulated in (15b). That is: What does 'replace' mean?

Let us turn to the notion of 'science' first. Quine, in his seminal paper (Quine 1969a), seems to hold the view that naturalism should be based on *behaviouristic psychology* and that 'science' is to be interpreted in the sense of *natural science*:

> [...] epistemology, or something like it, simply falls into place as a chapter of psychology and hence of *natural science*. It studies a natural phenomenon, viz., a physical human subject. (Quine 1969a: 82–83; emphasis added.)

However, as Haack (1993) showed, Quine is inclined to allow a considerably weaker interpretation as well, according to which 'science' must not be restricted simply to 'natural science' but should involve any kind of empirical inquiry such as the social sciences, linguistics etc. (cf. also Maffie 1995 on this). See for example the following illustrative quotation.

> What reality is like is *the business of scientists, in the broadest sense*, painstakingly to surmise; and what there is, what is real, is part of that question. [...] The last arbiter is *so-called* scientific method, however *amorphous* [...]. (Quine 1960: 22)

Accordingly, we obtain two possible explications of (15a).

(16) a. The naturalized philosophy of science claims to solve or to eliminate the problems which the analytic philosophy of science tackled on the basis of its own philosophical, a priori foundations, by using the a posteriori methods of *natural science.*

b. The naturalized philosophy of science claims to solve or to eliminate the problems which the analytic philosophy of science tackled on the basis of its own philosophical, a priori foundations, by using the a posteriori methods of any kind of empirical inquiry *not* restricted to natural science.

We will label the view represented by (16a) *strong naturalism* and the one put in (16b) *weak naturalism.* Now, let us introduce a *convention* so as to clarify the use of the term 'science':

(17) In what follows

a. *science* will refer to any kind of empirical inquiry including the social sciences, linguistics, history etc. in the sense outlined in Haack (1993), Maffie (1995);[25]

b. if we want to restrict the notion to the natural sciences, we will use the term *natural science*;[26]

c. in expressions like 'philosophy of science', 'cognitive science', 'cognitive science of science' the notion of 'science' will be used in the sense of (a).

In what different ways can we interpret the term 'replace' in (15b)? The answer that presents itself first is that the analytic philosophy of science should be eliminated *completely*. In this way Quine's famous claim that "epistemology is only science self-applied" (Quine 1969a: 293) means that traditional epistemology / the analytic philosophy of science should be radically discarded and the tasks of the latter should be adopted completely by science (whatever 'science' means).

However, Quine's writings are ambiguous with respect to the replacement of the analytic philosophy of science: in particular, he also assumes the continuity of science (in whatever sense) and philosophy (see also Haack 1993), which suggests that the latter *cannot* be replaced by the former completely. Thus we obtain the following two possible explications of (15b):

(18) a. The analytic philosophy of science should be *replaced completely* by 'science'.
b. The analytic philosophy of science should *not* be replaced completely by 'science'.

Consequently, we obtain two versions of strong naturalism depending on whether (16a) is combined with (18a) or (18b). Let me call *strong$_1$ naturalism* the set of theses {(16a), (18a)} and *strong$_2$ naturalism* the set {(16a), (18b)}.

The two versions of weak naturalism present themselves in an analogous manner. If we combine (16b) with (18a), we get *weak$_1$ naturalism*; and if we combine (16b) with (18b), the result is *weak$_2$ naturalism*. This very simple typology can be summarized in Table 1.1.

Table 1.1.

Versions of naturalism	Parameter 1: natural science as the only source of naturalized philosophy of science	Parameter 2: complete replacement of the analytic philosophy of science
strong$_1$	+	+
strong$_2$	+	–
weak$_1$	–	+
weak$_2$	–	–

Now the question arises as to the relationship between these four versions of the naturalized philosophy of science and cognitive semantics in the sense which we discussed in Section 1.2. We will give an answer to this question after one particularly important manifestation of the naturalized philosophy of science, namely, the cognitive science of science has been surveyed in the next section.

1.3.3 The cognitive science of science

As later developments showed, the scope of the naturalized philosohy of science could not be restricted to the metascientific use of natural science in general and empirical psychology in particular according to Quine's behaviouristic preferences. Rather, it was affected by the cognitive turn which emerged independently of the naturalistic turn in the philosophy of science and epistemology.[27] It is easy to see that the naturalized philosophy of science and cognitive science *share* a common background assumption: both strive, by using the methods of empirical disciplines, to investigate the way in which we gain knowledge of the world. The two trends seem to converge in that they maintain that it is both necessary and possible to reformulate traditional problems of epistemology as empirical problems to be solved by means of the empirical methods of science. The consequence is that now it is cognitive science which predominantly, though not exclusively, is expected to yield approaches to naturalized epistemology. The convergence of naturalized philosophy of science and cognitive science led to the emergence of a new field of research called the *cognitive science of science*: "much recent naturalized philosophy of science can be characterized as *cognitive science of science*" (Downes 1993: 453; emphasis added, A. K.). This also means that the naturalized philosophy of science, as manifested in the cognitive science of science, should radically change and widen the methods it uses. In contrast to Quine's original idea of behaviouristic psychology as the main source of naturalized epistemology/philosophy of science, the cognitive science of science is expected to comprise the subdisciplines of cognitive science which include fields traditionally labelled 'social sciences' (e.g. cognitive psychology) or 'humanities' (e.g. linguistics). In his survey of current trends in philosophy, J. Rouse summarized this situation as follows:[28]

> Philosophical naturalists have typically chosen a single scientific discipline as the grounds for explaining scientific research, and for most philosophers, that discipline has been psychology. Quine's influential [1969a] presumed that epistemology and philosophy of science should eventually be subsumed under *behavioristic* psychology (and perhaps ultimately neurophysiology). *As behaviorism has been increasingly supplanted by cognitive psychology, however, which in turn has been*

> *increasingly interactive with linguistics, computer science, and philosophy of mind, the predominant model for naturalistic philosophy of science has become interdisciplinary cognitive science.* (Rouse 1998: 89; emphasis added, A. K.)

This programme, of course, raises the question as to which particular methods the cognitive science of science should rest on. It is difficult to find an answer immediately, because, as we know, cognitive science is not a unified discipline, but rather, a system of subdisciplines which, though interacting with each other in an interdisciplinary manner, are governed by partially autonomous and independent factors. Should it be cognitive psychology, or cognitive sociology, or artificial intelligence research or neuroscience that serves as a starting point for the elaboration of the cognitive science of science? For example, Giere (1988) and Goldman (1986) focus on cognitive psychology; Langley et al. (1987), Thagard (1988), (1992) or Slezak (1989) put forward the idea of a computational science of science; Churchland (1986) and Churchland (1989) plead for neuroscience; Arbib and Hesse (1986) argue for the application of schema theory to the cognitive science of science. As an illustration, see for example Giere's summary of the situation:

> The major differences in approach among those who share a general cognitive approach to the study of science reflect *differences* in cognitive science itself. At present, 'cognitive science' is not a unified discipline, but an amalgam of parts of several previously existing fields, especially artificial intelligence, cognitive psychology, and cognitive neuroscience. *Linguistics*, anthropology, and philosophy also contribute. Which particular approach a person takes has typically been determined *more* by original training and later experiences than by the problem at hand. (Giere 2001a: 42–43; emphasis added)

This quotation serves to illustrate two important findings. The first concerns the fact that linguistics as a subdiscipline of cognitive science is one of the potential subdisciplines of the cognitive science of science as well. In Section 1.2.4 we have seen that the central subfield of cognitive linguistics, namely, cognitive semantics, exhibits the substantial features of the cognitive scientific enterprise in a clear and paradigmatic manner. Therefore, the following argument presents itself:[29]

(19) *If*
  a. cognitive semantics is one of the subdisciplines of cognitive science, and
  b. cognitive science is one possible manifestation of the naturalized philosophy of science called the cognitive science of science,

then

c. cognitive semantics is one possible manifestation of the naturalized philosophy of science called the cognitive science of science.[30]

Simple as the argument is, the conclusion in (19c) is of utmost relevance, because it opens *the prospect* of applying the methods of cognitive semantics to problems raised by the naturalized philosophy of science. Accordingly, cognitive semantics is one of the disciplines in which basic ideas of naturalism in general and those of the cognitive science of science in particular may be manifested. Consequently, due to the argument in (19) we have arrived at an important finding which we will call *the metascientific extension of cognitive semantics* (MECS):

(MECS) One of the tasks of cognitive semantics is to contribute to the solution of problems tackled by the naturalized philosophy of science in general and the cognitive science of science in particular.

The second point which the last quotation suggests concerns the concluding question of the previous section. In the light of (MECS) we may reformulate it more precisely: Which of the four basic types of naturalized philosophy of science is compatible with the metascientific extension of cognitive semantics (MECS)? The answer follows from the quotation immediately.

On the one hand, (a) cognitive science is an interdisciplinary enterprise, (b) the subdisciplines of cognitive science are not restricted to the natural sciences and (c) as witnessed by (12c) cognitive semantics itself cooperates with these subdisciplines in an interdisciplinary manner. Consequently, both strong versions have to be ruled out at the outset.

On the other hand, in the literature no convincing arguments have been put forward for the claim that the cognitive science of science should replace the analytic philosophy of science completely. Philosophy is a discipline which most subdisciplines of cognitive science interact with, therefore, it is plausible to assume that the cognitive science of science must not exclude it from the inventory of metascientific reflexion.[31] Consequently, after having ruled out both strong versions of naturalism, also weak$_1$ naturalism has to be abandoned. That is, the metascientifc extension of cognitive semantics *(MECS) is compatible only with weak$_2$ naturalism.*

This conclusion is important insofar as it legitimizes a substantially *pluralistic* approach to the metascientific application of cognitive semantics. Therefore, it follows that also from the point of view of the cognitive science of science it would be mistaken to confine our considerations to the prospects and limits of the metascientific extension of only one trend in cognitive semantics. If we want to explore the prospects and limits of cognitive semantics by extending it metascientifically, then we must proceed pluralistically on the metascientific level as well.[32] Thus, the plausibility of our decision to consider at least two incompatible trends, namely, modularism and holism simultaneously, is motivated not only by the state of the art in cognitive semantics, but also by the central characteristics of the cognitive science of science as well.

1.3.4 Conclusions

In this chapter we considered the question of whether, and if so, in what way current trends in the philosophy of science legitimize the use of cognitive semantic methods, if one sets out to capture certain aspects of scientific knowledge. As an answer, we obtained the metascientific extension of cognitive semantics (MECS). From this finding, as one possible specification of (Q′), we obtain (Q″):

(Q″) What are the prospects and limits of modular and holistic cognitive semantics *if* they are applied to scientific knowledge?

1.4 Summary

This means that we have to replace the very vague problem (Q) by the more specific problem (Q″). The latter narrows down the subject matter of our investigations to the prospects and limits of modular and holistic cognitive semantics, on the one hand, and to scientific knowledge, on the other.

Nevertheless, (Q″) is still too general to serve our purposes. One more necessary step is the selection of the particular cognitive semantic theories whose workability (Q″) focuses on. We will therefore specify (Q″) to two paradigmatic examples: *the cognitive theory of metaphor* as the representative of holism and *the two-level approach* as a paradigmatic example of modular cognitive semantics. The reasons for choosing these two approaches are as follows:

- Both are relatively clear cases of modularism or holism; they explicitly show all the properties of the particular trend which were enumerated in (10a–h) and (11a–i) in Section 1.2.

- Both are extremely successful manifestations of modularism or holism. 'Successful' here means that since the beginning of the eighties they have been the focus of interest in at least two respects: they have been widely *applied*, and they have been constantly *criticized*.
- With respect to the *cognitive theory of metaphor* this means that the latter is perhaps the most popular trend within cognitive semantics which has been discussed and refined in numerous publications and which played an important role in making some of the ideas of cognitive linguistics widely known; at the same time, as is usually the case with influential achievements, it was exposed not only to enthusiastic appraisals but also to fierce rejection and hard criticism as well.
- The current status of the *two-level approach* can be properly characterized as follows.

> Bierwisch (1983) [i.e. 1983b, A. K.] made a *seminal* contribution to the rise of a semantic conception which (at least in the German speaking research community in the area of lexical semantics) has been in intensive discussion thereafter. The theory is known as the two-level approach although Bierwisch himself has never given it a name. Many of the questions he asked in his paper are *still unanswered.* And some of the solutions he proposed are *still a matter of intensive discussion.* The main thesis of his approach is the strict distinction between a linguistic-semantic level of meaning, and an essentially non-linguistic conceptual level at which the semantic representations are interpreted, relative to conceptual knowledge of the world. (Börkel 1995: 69)

In the next chapter these two theories will be *introduced* and *developed into* cognitive semantic approaches to the cognitive science of science.

CHAPTER 2

Two metascientific extensions of cognitive semantics

2.1 Introduction

In Section 1.3.3 we put forward the metascientific extension of cognitive semantics (MECS), and (MECS), in turn, yielded (Q″) in Section 1.3.4. We suggested the two-level approach to cognitive semantics and the cognitive theory of metaphor as paradigmatic examples which should illustrate the workability of this thesis. In the light of (Q″), it is therefore important to realize that both the proponents of modular cognitive semantics and those of holism consider the metascientific extension of their particular approach to be *fully legitimate*. Bierwisch, for example, points out the following in one of the classic contributions to the two-level approach.

> [Organizational principles] provide the framework according to which *common sense as well as scientific knowledge* develops.
> These principles play different roles with respect to different concepts, and they may allow, moreover, for different stages in the development of a given concept. Water for instance might first be characterized as to structure and function, and only later on by something like H_2O for substance. Instead by a fixed and uniform stereotype, a concept will thus be represented by a flexible schema of distinctions emerging from the principles in terms of which experience is accommodated. On this account, *common sense and expert concepts* are not necessarily separated, but are different, though compatible differentiations with respect to alternative principles. A crucial assumption underlying these considerations is the interdependence of concepts. *Common sense explanations, just as scientific theories of different kinds*, are not collections of isolated concepts, but rather connected systems that organize coherent domains of experience. (Bierwisch 1983a: 64–65; emphasis added)

It is also emphasized that this framework is capable of capturing the nature of theoretical terms in linguistics itself. Illustrating one of the conceptual operations he introduced, namely, what he calls 'conceptual shift' and the related notion of 'family of concepts' (see Section 2.2.1 for an explanation of these and Chapter 5 for their application to the theoretical terms of generative linguistic theory), Bierwisch writes:

> Further types of lexically generated *families of concepts* can easily be added. Consider for example the particularly intriguing families determined by word, phrase, or language. It is furthermore to be noticed that conceptual distinctions of the type in question frequently remain completely implicit, escaping any explicit discrimination, until *particular demands come up.* In fact, the recent history of linguistics consists to a reasonable extent in the *clarification of the concepts* associated with language, eventually distinguishing them terminologically by competence, performance, dialect, idiolect, communication, etc. (Bierwisch 1983a: 67; italics added)

Nevertheless, these remarks refer to a marginal rather than to a central feature of the two-level approach. Although, as witnessed by the above quotations, scientific concept formation is considered to be one of its intended applications, such applications were never carried out explicitly by its proponents.[1]

Lakoff and Johnson highlight the applicability of their cognitive theory of metaphor to expert discourse. To mention only two illustrative examples, they characterize certain properties of scientific theories in this way:[2]

> So-called purely intellectual concepts, e. g. the concepts in a scientific theory, are often — perhaps always — based on metaphors that have a physical and/or cultural basis. The *high* in 'high-energy particles' is based on MORE IS UP. The *high* in 'high-level functions', as in physiological psychology, is based on RATIONAL IS UP. The *low* in 'low-level phonology' (which refers to detailed phonetic aspects of the sound systems of languages) is based on MUNDANE REALITY IS DOWN (as in 'down to earth'). The intuitive appeal of a scientific theory has to do with how well its metaphors fit one's experience. (Lakoff and Johnson 1980a: 19; emphasis original)

> Metaphor is the main mechanism through which we comprehend abstract concepts and perform abstract reasoning. Much subject matter, from the most mundane to the most abstruse *scientific theories*, can only be comprehended via metaphor. (Lakoff 1993: 244; emphasis added).

In contrast to the two-level approach, the cognitive theory of metaphor has been applied widely to scientific knowledge. In later publications such as Lakoff and Johnson (1999), and Lakoff and Núñez (2002), the authors explicitly extend the scope of the cognitive theory of metaphor to scientific, philosophical and mathematical concept formation. Important applications of the cognitive theory of metaphor to the analysis of certain aspects of scientific knowledge have been carried out for instance in Baldauf (1997), Jäkel (1997), Kövecses (2000) and Drewer (2003).

Thus we have two approaches at our disposal which differ, among other things, in two respects. Firstly, as we have seen in (10) and (11) in Chapter 1, one of them accepts the modularity hypothesis, whereas the other maintains the holistic hypothesis. Secondly, although both emphasize that scientific concept-formation belongs to their intended applications, it is only the cognitive theory of metaphor which carried out such investigations extensively. However, this doesn't speak against testing the two-level approach with respect to scientific concept formation, too. Rather, examining one of its hidden properties by manipulating it in the course of a new application is in full accordance with the nature of thought experiments discussed briefly in Section 3 of the Introduction. It is exactly this kind of manipulation of the model which is expected to reveal how far its scope can be extended, that is, what some of its prospects and limits may be.

So as to clarify the immediate consequences of the intended applications of the cognitive theory of metaphor and the two-level model mentioned in the quotations, we have to introduce their basic notions and their basic tenets systematically. In Sections 2.2.1 and 2.3.1 we will restrict the presentation of the two theories to some basic aspects of their *classic versions* so as to provide a coherent basis for our argumentation. If necessary, later developments of the theories will be considered with respect to the particular problems raised in subsequent chapters. Moreover, we do not strive to give a systematic comparison of the two approaches by contrasting their main theses, for establishing the criteria of such a comparison would raise substantial difficulties. Rather, we will try to present their main tenets in a way which is as close to the original formulations as possible. In Sections 2.2.2 and 2.3.2 the metascientific application of the two approaches will be motivated. As a result, Section 2.4 will narrow down (Q″) to (Q‴) which will be regarded as the main problem of the book.

2.2 A paradigmatic example of modular cognitive semantics

2.2.1 The basic tenets

The basic assumption is the *modularity hypothesis* introduced in Section 1.2.3. which we will label (MH):[3]

(MH) Knowledge of language is a relatively autonomous system (=module) interacting with other systems and is itself structured by the interaction of relatively autonomous (sub)systems.

The two-level approach rests on the generalization of (MH):[4] "Basically, *all human cognitive behaviour is organized in a modular fashion.*" (Lang and Carstensen 1990: 6; emphasis added, A. K.) Thus *the generalized modularity hypothesis* is as follows.

(MH′) Human cognitive behaviour is organized in a modular way.

(MH′) is related to a series of assumptions which characterize the nature of modules within the framework of the two-level approach. The following tenets are specifications of the general assumptions of modularism enumerated in (11a–i) in Section 1.2.3.[5]

(1) At present there is very little evidence at our disposal for identifying the modules of human behaviour. Nevertheless, hypothetically, at least the following ones may be assumed (see Bierwisch 1981, Grewendorf et al. 1987, Lang and Carstensen 1990 for these possible candidates):

(i) a. The *motoric module* governs the functioning of human organs and is responsible, for instance, for the articulation of sounds, for gestures, for mimicry etc.
b. The *module of perception* underlies processes of human perception and comprises submodules like that of vision, hearing etc.
c. The *conceptual module* structures the mental representations of human experience.
d. The *module of social relations* comprises at least the following two submodules:
   – the *submodule of social interactions*, and
   – the *motivational submodule* which organizes the objectives, interests and intentions of individuals and groups of individuals.
e. The *grammatical module* consists of the phonological, the morpho-syntactic and the semantic subsystem.

However vague and hypothetical the assumptions concerning the modules are, basically the two-level approach presupposes the computer-metaphor according to which "the mind is a computer" (Bierwisch 1987, 1990). Thus, the two-level approach has been integrated with artificial intelligence research and has motivated the development of computer programmes modelling knowledge representation (see Lang 1991, Lang and Carstensen 1990, Lang et al. 1991).

(2) *Instances* of cognitive behaviour consist of sets of *representations*. These representations are determined by *rules* and rules, in turn, are determined by *principles* which may be universal or specific to a certain module.[6]

(3) By definition, every module is *relatively autonomous* (cf. (11b) in Section 1.2.3.). The relative autonomy of modules is manifested in the fact that the

universal principles which they involve are associated with *free parameters*.[7] The important claim is that the value of a parameter P_1 associated with a universal principle UP_1 in the module M_1 may 'depend' on the value of a parameter P_2 associated with a principle UP_2 belonging to the module M_2. This relation between UP_1 and UP_2 is called the *parametrization relation* (Bierwisch 1981, Lang and Carstensen 1990, Lang et al. 1991).

(4) The universal principles determine particular rules of cognitive behaviour. From the way the parametrization relation has been formulated, it therefore follows that a particular rule is nothing but a principle whose *parameter* has been *fixed* at a certain value. If the parameter is fixed at different values in different cases, then we get different rules. The second consequence is that the values of the parameters which in fact yield specific rules may depend on the values of parameters belonging to another module and therefore, a rule R_1 belonging to a module M_1 may not simply be determined by the principles of this module M_1 only, but can also result from the parametrization relation between at least two universal principles UP_1 and UP_2 belonging to two different modules M_1 and M_2, respectively (Bierwisch 1981, Lang and Carstensen 1990, Lang et al. 1991 etc.).[8] These rules, in turn, determine particular representations R_i via the parametrization relation (see also 3).

(5) A linguistic utterance is conceived of as an 'inscription' *ins* and a 'linguistic structure' *ls*. The latter is a triple of a 'phonetic', a 'morphosyntactic', and a 'semantic' representation *pt, syn, sem,* respectively (Bierwisch 1983a, 1983b etc.):[9]

(i) $u = <ins, <pt, syn, sem>>$

One important remark concerns the morphosyntactic representation *syn*. Namely, the morphosyntactic representation of linguistic utterances may be described by some version of generative syntax.

The second remark should emphasize that the semantic representation *sem* is one of the central constructs of the two-level approach. In fact the considerations put forward within the two-level approach focus on the nature of *sem* in a substantial manner and try to explore its main aspects. As a first approximation it is assumed that *sem* determines an 'utterance meaning' *m* by the help of a 'context' *ct* with respect to which *u* is interpreted. Thereby *ct* and *m* belong to the conceptual module; the context is the conceptual environment of a certain utternce and need not be verbalized. *Sem* belongs to the grammatical module while, at the same time, it works as the interface between the grammatical and the conceptual module: "[…] the semantic representations mediate between conceptual representations and the combinatorial structure of syntax" (Bierwisch 1983a: 69). Thus we get this:

(ii) $\langle\underbrace{\langle pt, syn, sem\rangle}_{G}, \underbrace{ct, m}_{C}\rangle$

Here *G* and *C* stand for the grammatical and the conceptual module, respectively.

(6) There is, among other things, a clear-cut intuition behind the above notions: namely, that modular cognitive semantics may adapt the technique of formal semantics in that the 'empty' set theoretical structures of the latter are related to 'mental states' i.e. some of the modules mentioned in (1) above. This means that an essential *move is carried out from formal to modular cognitive semantics which facilitates their continuity*. From this point of view it is central that what formal semantics calls 'extension', 'intension' and 'possible worlds' may be retained, but should be related to certain modules. Then, *sem* is what in formal semantics is called 'intension' whereas *m* and *ct* belong to the extension of *u*: "There are two structures to be specified: The domain of extensions, i.e. of utterance meanings and contexts, and the system of intensions, i.e. of semantic representations" (Bierwisch 1983a: 43, see also Bierwisch 1983b: 79, etc.).[10]

(7) So as to make this move comprehensible, let us continue the introduction of the basic categories of the two-level approach that we began in (5).

A triple consisting of an utterance *u*, its utterance meaning *m*, and the context *ct* is a 'meaningful utterance' *mu*:

(i) *mu* = *<ins, <<pt, syn, sem>, ct, m>>*

What has been defined so far belongs to the domain of 'language', and involves two modules, namely, the grammatical and the conceptual. Turning now to the domain of 'social interaction', a 'speech act' may be defined as a meaningful utterance to which a 'communicative sense' *cs* is assigned relative to a certain 'interactional setting' *ias*.[11] That is:

(ii) *sa* = *<ins, <<<pt, syn,sem>, ct, m>, ias, cs>>*

ias and *cs* are representations belonging to the module of social interaction (*I* stands for the interactional module):

(ii) $$\langle\langle\langle\underbrace{\langle pt, syn, sem\rangle}_{G}, \underbrace{ct, m\rangle}_{C}, \underbrace{ias, cs\rangle\rangle}_{I}$$

Therefore, *sa* is to be conceived of as a representation belonging to the interactional module, too.[12]

(8) These notions also indicate that another important move which the two-level approach carried out is one which leads from the conception of 'knowledge of language' as a mental system to its treatment *both as a mental system and as a social phenomenon.*[13]

(9) *Lexical semantics* is of special interest for the two-level approach. The latter focuses on cases of polysemy which can be captured neither as ambiguity nor as vagueness and are restricted to literal meaning.[14] The next paragraphs will elucidate this kind of polysemy and the way the two-level approach accounts for it.

(10) One of the techniques which the two-level approach makes use of so as to capture word meaning is the integration of *intensional semantics* and *componential semantics.*[15]

(11) The interpretation of *lexical items* is specified as follows (see e.g. Bierwisch 1983a, 1983b, Lang 1994: 26 etc. for details):

(i) *sem(A, ct)=m.*

Here *A* is a lexical item, *ct* is the context, *m* is the conceptual representation of *A* and *sem* is the semantic representation of *A*. *Sem* is, as already mentioned, the interface between the grammatical and the conceptual module, whereas *ct* and *m* belong to the latter. A very important aspect of (i) is that, as a result of the intermodular definition of interpretation, what may be called the 'meaning' of a term is *semantically underdetermined*, because it is not only semantic information, but also information from the conceptual module which it makes use of: "the notion of meaning refers in part to the semantic structure of a word, and in part to the conceptual unit expressed by a word in a given utterance [...]" (Bierwisch 1983a: 89). With respect to later considerations it is important to remark that the context *ct* is to be understood as a *free parameter* which may be fixed at different values; accordingly, the value of *m* depends on the value at which *ct* is fixed (Bierwisch 1983b: 79).

(12) We may assume that a lexical item *A* is associated with a *family of concepts*.[16] Such a family of concepts associated with a particular lexical item may be characterized in this way:

(i) a. a *family of concepts FC* is a set of conceptual representations of the lexical item *A*, i. e. $FC = \{m_j\}$;
b. any element of this set is determined by a function *V* such that $V(sem, ct_i) = m_j$.

The function *V* selects from the set of conceptual representations mentioned in (ia) relative to a context ct_i one of these conceptual representations as the interpretation of *A* in ct_i. In accordance with the generalized modularity hypothesis (MH′) we have to assume that the modules consist of universal principles and rules. Consequently, this applies to the conceptual module as well, and this in turn means that the relationship between the representations determined by the rules and principles of the conceptual module is based on such principles and rules. (Cf. Bierwisch 1983a, 1983b, Lang and Carstensen 1990, Lang et al. 1991 etc. with respect to (MH′).) Thus, according to Bierwisch (1983a: 93) "*V* captures relevant generalizations concerning the conceptual interpretation of semantic structures. [...] *V* is based on a system of principles and rules comprising among others [...]" *conceptual shift* and *conceptual specification.*

(13) In particular, *conceptual shift* means that the interpretation of *A* is 'shifted' from one conceptual area to another via *sem.* For example (see Bierwisch 1983a: 66 f.):

(i) *I put the <u>letter</u> on your desk.*
(ii) *The <u>letter</u> has been distributed to the whole faculty.*
(iii) *The <u>letter</u> finally led to a political crisis.*
(iv) *For many poets, the <u>letter</u> is a genuine literary genre.*

In (i) the word *letter* represents a physical object of a certain kind; in (ii) it represents a set of such objects; in (iii) it stands for the informational content; finally, in (iv) it represents the type of informational structure. The idea is that in all these cases the interpretations of *letter* are, on the one hand, closely related to each other, and yet on the other hand, they are different conceptual representations of the same lexical item depending on different 'conceptual domains'.[17] This means that a lexical item is related to an organized family of concepts, while all the members of this family are connected to the same semantic representation *sem*.

This example also shows in what sense the semantic underdetermination which underlies the particular interpretations of *letter* in (i)–(iv) is to be distinguished both from ambiguity and vagueness. The word *letter* is not ambiguous, because it is not the case that the *semantics* of this lexical item includes several interpretations specified in the lexicon at the outset. And it is not vague, either, because it is not the case that we cannot draw the dividing line between two interpretations. Rather, we have to do with a kind of polysemy characterized by the fact that the differences between the interpretations of *letter* in (i)–(iv) seem to result from the interplay of the context and some of the semantic properties of the lexical item, while these properties alone are not sufficient for determining the particular interpretations.

(14) A second principle is *conceptual specification*, the main characteristic of which is that *sem* is specified according to the contexts, but — unlike in the case of conceptual shift — within the same conceptual domain, without being shifted to different 'conceptual domains'. In one of his early works Bierwisch (1983a: 68) introduces this principle by the example of *lose*:

(i) *John <u>lost</u> his money, as he was not aware of the hole in his pocket.*
(ii) *John <u>lost</u> his money by speculating at the stock market.*
(iii) *John <u>lost</u> his friend in the overcrowded subway station.*
(iv) *John <u>lost</u> his friend, as he could never suppress bad jokes about him.*

Just as *letter* in the examples in (13), also *lose* is neither vague nor ambiguous. Bierwisch analyzes the semantic representation *sem* of the lexical item *lose* in the following way.[18]

(v) λX[λY[∃Z[CAUSE Z [GO-TO[NOT[HAVE Y X]]]]]]

The point is that *Z* is a variable which ranges over events to be specified at the conceptual level. We get the particular conceptual interpretations from (v) in that this free variable is bound by an existential quantifier. The quantifier specifies those particular elements in the context of interpretation which can be inserted in (v) and which yield particular conceptual representations of (v).

Each value of the variable *Z* within the context of interpretation *ct* yields a particular conceptual representation m_i (i.e. interpretation) of the expression *lose.* As the examples in (i)–(vi) illustrate, "conceptual specification does not create different types or categories of concepts, but rather, various specifications within one and the same conceptual type by filling in open slots, so to speak" (Bierwisch 1983a: 68).

(15) At this point the mechanisms we have sketched should be further illuminated by some additional comments:

(a) Having illustrated the two conceptual principles mentioned by simple examples, now we may explicate the idea of the semantic underdetermination (introduced in (11)) of lexical items in the following way:

(i) A lexical item *A* is semantically underdetermined if there is a function *V′* such that
- *V′* maps the semantic representation *sem* of *A* onto a family of concepts *FC*, and
- *FC* contains more than one conceptual representation *m* as its elements.

(b) In interpreting one and the same lexical item, the two operations mentioned can be combined (Bierwisch 1983b: 92) and their combination yields a complex network of conceptual representations and conceptual operations.

(c) There are considerable differences in the degree of semantic underdetermination concerning particular lexical items. On the one end of the scale there are lexical items which don't allow any variability in their interpretations at all or allow such a variability only to a restricted extent. On the other end of the scale we find items which cannot be interpreted at all without assuming the rich variability of their interpretations.[19] For example, *eleven* shows only restricted variability, whereas *have* cannot be interpreted at all without assuming the strong semantic underdetermination that is the rich variability of its interpretations. See Bierwisch (1983b: 92).

(d) Finally, there is also a third principle which appears to be relevant: *conceptual selection.* It operates when terms which are structured by one of the principles mentioned above — i.e. conceptual shift and/or conceptual selection — are coordinated, as a result of which the interpretation of one term selects the interpretation of another (Bierwisch 1983a, 1983b). For example, in (14i) *money* refers to concrete objects and *lose* refers to a change of location; or, in (ii) *money* is to be understood as an abstract exchange value and *lose* as a change of possession etc.

(16) It may be useful to make a short digression to consider the question of how the two-level approach handles metaphors. As already mentioned in (11i) in Section 1.2.3, the two-level approach does not focus on metaphors. In fact, the three types of conceptual operations mentioned, which the two-level approach is primarily interested in, are restricted to literal meaning. Nevertheless, Bierwisch (1979) considers the principle of *metaphorization* as well. In this often-cited paper, metaphors are treated as follows. Let us assume that the 'literal meaning' *LM* of a term *A* is determined by the function $sem(A, ct_a)$, where ct_a is a context characterized by the fact that it does not contain pieces of information inconsistent with *sem.*[20] What matters, is that Bierwisch infers the 'metaphorical meaning' *MM* of a term from its 'literal meaning' *LM.* Although metaphorical meanings are always established in 'non-neutral' contexts, there is for every non-neutral context ct_a' a nearest neutral context ct_a which determines the 'literal meaning' *LM.* The metaphorical meaning *MM* of a term *A* is yielded by the conceptual principle $M(LM, ct_a)$. The effect of this principle is that in the non-neutral context ct_a' the metaphorical meaning *MM* is defined as the minimal change of the literal meaning *LM* of *A* consistent with the latter. It is also important to remark that under specific circumstances the principle of metaphorization may interact with the operations illustrated above which are normally restriced to the literal use of expressions.[21] Bierwisch (1979: 143) emphasizes that there is no simple sequentiality between the literal and the non-literal interpretations of expressions; we must not assume mechanically that the metaphorical interpretation has always to follow the literal. Rather, a complex and sophisticated interplay between 'literal' and 'non-literal' interpretations has to be assumed, whose particular structure may differ from case to case.

(17) Finally, the important methodological remark has to be made that the two-level approach

- conforms to the tradition of main-trend Anglo-American analytical philosophy (see e.g. both the use of formal semantics and the acceptance of basic ideas of speech act theory);
- converts some of the main issues of the latter into empirical problems and empirical solutions to these problems along the lines we discussed in

Section 1.2.1 (see e.g. the interpretation of the formalism of semantics as mental states);
- conforms to those trends within cognitive science which are compatible with Chomskyan generative linguistics;
- closely interacts with artificial intelligence research.

Accordingly, the two level-approach strives to *integrate* the following methods:

- formal semantics (see (6)),
- componential semantics (see (10)),
- speech act theory (see (7))
- generative syntax (see (5)).
- artificial intelligence research (see (1)).
- possible approaches to social interactions.

Thus the two-level approach is *much more than a semantic theory*: rather, it is an *integrated approach to human cognitive behaviour* which is intended to capture the complexity of the latter by allowing the interaction of different theoretical accounts. The coherent integration of these tools as subtheories of the two-level approach is facilitated by their *mentalistic reinterpretation.*[22] Moreover, the above survey also shows that Harder's characterization quoted in Section 1 of the Introduction according to which in cognitive linguistics in general and cognitive semantics in particular "one can move freely and gradually from facts about *language* to facts about *human cognition* and further on to facts about *human life generally* [...]" (Harder 1999: 196; emphasis added) applies not only to holism, but to the two-level approach, too.

2.2.2 The metascientific extension

If we relate an empirical hypothesis like (MH') to the metascientific extension of cognitive semantics (MECS) introduced in Section 1.3.3, then (MH′) may be transferred to the metascientific level. Thus, a very simple plausible argument presents itself:[23]

(18) *If*
  a. 'scientific knowledge' is part of human cognitive behaviour, and
  b. human cognitive behaviour is organized in a modular way along the lines of the generalized modularity hypothesis (MH′),

  then

c. we obtain *the metascientific extension of the generalized modularity hypothesis*:

(MMH′) 'Scientific knowledge' is organized in a modular way along the lines of the generalized modularity hypothesis (MH').

This means that 'scientific knowledge' is based on the interaction of modules.[24] Consequently, the assumptions specified and illustrated in (1)–(17) apply to scientific knowledge as well and constitute, in accordance with the metascientific extension of cognitive semantics (MECS), *a possible approach to the cognitive science of science.*

2.3 A paradigmatic example of holistic cognitive semantics

2.3.1 The basic tenets

For the sake of argument, in a very simplified manner which, however, emphasizes the contrast between the modular and the holistic approach, the *holistic answer to question (9)* in Section 1.2.3 can be generalized so that we obtain *the generalized holistic hypothesis*:

(HH) Human cognitive behaviour constitutes a unified system.

This reformulation of the holistic hypothesis is in full accordance with (10) in Section 1.2.3. Admittedly, one of the reasons why this and similar reformulations are legitimate is the constitutive terminological vagueness of cognitive semantic theories which is something we cannot change. 'Human cognitive behaviour' is assumed to comprise 'Knowledge of language and other cognitive systems' mentioned in (10) in Section 1.2.3. Nevertheless, there is another important motivation for reformulating the latter as (HH) which is of a heuristric nature: in particular, replacing 'knowledge of language and other cognitive systems' with 'human cognitive behaviour' allows an analogous treatment of the holistic hypothesis with the generalized modularity hypothesis from the point of view of argumentation.

With respect to *the cognitive theory of metaphor* as elaborated by Lakoff and Johnson, the specific manifestation of this general assumption is the claim that "[...] the human conceptual system is metaphorically structured [...]" (Lakoff and Johnson 1980a: 6). Thus our reconstruction of *the main hypothesis of the cognitive theory of metaphor is (HH′).*[25]

(HH′) Human cognitive behaviour is structured metaphorically.

(HH′), in turn, may be spelled out in the form of a set of even more specific assumptions.[26]

(19) Metaphor is not a rhetorical phenomenon restricted to poetic language, but rather, it is constitutive both of everyday language and expert discourse (Lakoff and Johnson 1980a: 3). It is "a pervasive, indispensable structure of human understanding by means of which we figuratively comprehend our world" (Johnson 1987: XX). Metaphors are to be considered to belong to linguistic competence in general.[27] Moreover, metaphors are *conceptual* entities and the conceptualization of reality is basically metaphorical — metaphors play a central role in cognition. Metaphor is "a matter of central concern, perhaps the key to giving an adequate account of understanding." (Lakoff and Johnson 1980a: IX).

(20) There is a substancial difference between *metaphorical expressions* and what Lakoff and Johnson call *metaphorical concepts;* they also use the expression 'conceptual metaphor' which is considered to be synonymous with 'metaphorical concept'. The relationship between metaphorical expressions and metaphorical concepts may be characterized by at least three important aspects:

a. *Metaphorical expressions* are the linguistic manifestation of *metaphorical concepts*: "Metaphors as linguistic expressions are possible precisely because there are *metaphors in a person's conceptual system*" (Lakoff and Johnson 1980a: 6; emphasis added). Metaphorical concepts fulfill all their functions not individually, but through a network of metaphorical expressions.[28] By way of illustration, let us take the following example:

Conceptual metaphor:	LOVE IS A JOURNEY
Metaphorical expressions:	*this relationship is foundering,* *we are going nowhere,* *this relationship is a dead-end street, we are at a crossroads,* etc.

b. It is important to remark that whereas metaphorical concepts (=conceptual metaphors) are primary with respect to metaphorical expressions in that the latter are manifestations of the former, *methodologically* the cognitive *theory* of metaphor reconstructs metaphorical concepts on the basis of metaphorical expressions:

> Since metaphorical expressions in our language are tied to metaphorical concepts in a systematic way, *we can use metaphorical linguistic expressions to study the nature of metaphorical concepts* and to gain an understanding of the metaphorical structure of our activities. (Lakoff and Johnson 1980a: 7; emphasis added).

c. Metaphorical expressions are not isolated phenomena, but rather, they constitute a *network* of expressions which are the linguistic manifestations of conceptual metaphors. This means, among other things, that the notion of conceptual metaphor facilitates *generalizations* which without this notion would not be possible: "Without conceptual metaphor a large range of generalizations cannot be stated" (Lakoff 1990: 42).

(21) Metaphors connect two conceptual domains: the *target domain* and the *source domain*. In the course of metaphorical processes the source domain *corresponds* to the target domain; in other words, there is a *mapping* or a *projection* between the source domain and the target domain. The target domain *X* is understood in terms of the source domain *Y*. For example, in the case of the metaphorical concept mentioned above, LOVE is the target domain whereas JOURNEY is the source domain. Whenever JOURNEY is mapped onto LOVE, the two domains correspond to each other in a way which enables us to interpret LOVE as JOURNEY.[29]

(22) The source domain is based on sensory experience and is *concrete*, while the target domain is *abstract*.

(23) The *Thesis of Unidirectionality* is one of the central tenets of the cognitive theory of metaphor.[30] The main idea is that the metaphorical projection takes places in one direction only, namely, from a concrete source domain towards an abstract target domain — in this sense it is irreversible.

(24) In the course of metaphorical projection, certain patterns inherent in the source domain are projected onto the target domain. Since it is hypothesized that these image schemas preserve their invariance during the projection, it follows that even very abstract domains are rooted in sensory experience.[31]

(25) Consequently, one of the main properties of metaphorical concepts is their *explanatory role*. This means that we capture abstract domains and theoretical constructs by conceptualizing them along the lines of metaphorical projection. As a result of the fact that abstract domains are traced back to

sensory experiences via metaphors, the former are explained in terms of the latter. Accordingly, right at the outset it is assumed that metaphorical concepts yield, among others, 'theoretical constructs' or 'theoretical terms' which assume central positions in scientific theories (see also Jäkel 1997: 42).[32]

(26) Metaphors also lead to the restructuring of conceptual domains and can, accordingly, *create new realities* (Lakoff and Johnson 1980a: 146, 211).

(27) However, conceptual metaphors capture the target domain *only partially* in that they focus on certain aspects and blur others. Thus, they fulfill a specific kind of *focussing function* (Jäkel 1997: 37).[33]

(28) Metaphors are closely connected to *idealized cognitive models* of certain segments of reality:

> [...] we organize our knowledge by means of structures called idealized cognitive models, or ICMs, and category structures and prototype effects are by-products of that organization. [...] Each ICM is a complex structured whole, a gestalt, which uses four kinds of structuring principles: [...] metaphoric mappings, as described by Lakoff and Johnson [...]. (Lakoff 1987: 68)

(29) These cognitive models constitute *scenarios*:

> A scenario consists fundamentally of the following ontology: an initial state, a sequence of events, and a final state. In other words, the scenario is structured by a source-path-goal schema in the time domain, where
>
> – the initial state = the source
> – the final state = the destination
> – the events = location on the path
>
> and the path stretches through time. The scenario is a WHOLE and each of these elements is a part.
>
> The scenario ontology also consists typically of people, things, properties, relations, and propositions. In addition, there are typically relations of certain kinds holding among the elements of the ontology: causal relations, identity relations, etc. These are represented structurally by link schemas, each of which is categorized as to the kind of relation it represents. Scenarios also have a purpose structure, which specifies the purposes of people in the scenario. Such structures are represented metaphorically via SOURCE-PATH-GOAL schemas, as discussed above. (Lakoff 1987: 285–6)

(30) *Embodiment* and experience are key notions of the theory at issue:

> Conceptual embodiment: The idea that the properties of certain categories are a consequence of the nature of human biological capacities and of the *experience* of functioning in a physical and social environment. It is contrasted with the idea that concepts exist independent of the bodily nature of any thinking beings and independent of their experience. [...]
> Cognitive models are directly embodied with respect to their content, or else they are systematically linked to directly embodied models. Cognitive models structure thought and are used in forming categories and in reasoning. Concepts characterized by cognitive models are understood via the embodiment of the models. (Lakoff 1987: 12–13; emphasis by underlining in the original)

In accordance with this, the cognitive theory of metaphor is also called 'experientialism' or 'embodied realism'

(31) Although we have deliberately avoided a systematic comparison of the theses of the two-level approach and those of the cognitive theory of metaphor, at this stage it is both possible and instructive to contrast briefly at least the methodological self-evaluation of the latter with some main properties of the former as summarized in (17). In particular, the cognitive theory of metaphor

- rejects the tradition of Anglo-American analytical philosophy;
- converts some of the central problems of classic philosophy into empirical problems by solving them on the basis of assumptions which revolve around the metaphorical structure of concept formation and the embodiment of conceptual models;
- rejects those trends within cognitive science which are compatible with generative linguistics.[34]

2.3.2 The metascientific extension

In accordance with the ideas underlying the cognitive science of science outlined in Section 1.3.3, the main hypothesis of the cognitive theory of metaphor (HH′) and the metascientific extension of cognitive semantics (MECS) yield the following:

(MHH′) 'Scientific knowledge' is structured by metaphorical concepts along the lines of the main hypothesis of the cognitive theory of metaphor (HH′).

This thesis will be called *the metascientific extension of the main hypothesis of the cognitive theory of metaphor.* As a consequence of (MHH′), all the tenets enumerated in (19)–(31) apply to scientific knowledge as well. Therefore, due to (MHH′), the cognitive theory of metaphor may be regarded as *a possible approach to the cognitive science of science.*

2.4 Summary

(MHH′) and (MMH′) outline two alternative approaches to the cognitive science of science. In this way we may specify the problem (Q″), which was put forward in 1.3.4., as (Q‴):

(Q‴) What are the prospects and limits
a. of the metascientific extension of the main hypothesis of the cognitive theory of metaphor (MHH′),
and
b. of the metascientific extension of the generalized modularity hypothesis (MMH′)?

This question seems to be precise enough to be subdivided into minor problems which will be the subject matter of case studies whose task is to exemplify certain prospects and limits of cognitive semantics as applied to certain aspects of scientific knowledge. Therefore, *we will consider (Q‴) as the main problem, and its solution as the main objective of the present work.*

In Part II, case studies will be developed that illustrate *the prospects* of the two frameworks based on (MHH′) and (MMH′), respectively. These case studies will center on one of the classic problems of the philosophy of science: the so-called problem of theoretical terms. It is no accident that the power of the two approaches will be tested by examining their capacity to handle this problem. The first reason is that both the two-level approach and the cognitive theory of metaphor focus, among other things, on lexical semantics, that is, on the 'meaning' of what are also called 'terms'. Therefore, it is quite straightforward to choose a problem in the philosophy of science whose possible solution is related to the semantics of lexical units at the outset. The second reason is that the problem of theoretical terms is one of the most fundamental quandaries of the philosophy of science which has been the focus of attention for a couple of decades and which has shaped the course of its development to a considerable extent. Thus, it is very well suited for testing the power of metascientific approaches.

Part II

Prospects: Theoretical terms

Chapter 3

The background

3.1 Introduction

One of the most important factors which the workability of scientific theories hinges on is the fact that they introduce, among other things, so-called *theoretical terms.* Well-known examples of such theoretical terms are *atom*, *electron*, *mass*, *force*, etc. They are applied in scientific generalizations which are indispensable for the *explanation* of facts. Moreover, the most frequent *method* of introducing theoretical terms into a theory is *explication.* So, we are justified in turning briefly to the relationship between the notions 'scientific explanation', 'explication' and 'theoretical terms' in the philosophy of science.[1]

3.2 On the problem of theoretical terms

The role which theoretical terms play in scientific investigation can be characterized by an interesting dichotomy. On the one hand, as indicated, they are *conducive to* the construction of scientific theories, to the explanation of the facts observed, and to our scientific understanding of the world in general. On the other hand, their status is still highly *questionable*, because we do not know by what mechanism these important processes are carried out: in particular, however strange it may seem at first sight, although they refer to 'unobservable' objects, it is theoretical terms which establish a *relationship* between theory and 'observable' reality. Therefore, one of the central topics of the analytic philosophy of science is *the problem of theoretical terms*, namely, the question of how this dichotomy can be captured. The development of the most important views within the analytic philosophy of science regarding this problem can be summarized in the following way.

(1) The first influential attempt by Hempel, Carnap, and others, proposed a distinction between theoretical and observational terms: unlike theoretical terms, observational terms were assumed to refer clearly to observable objects.

It was assumed that theoretical terms could be reduced completely to observational terms. In particular, this meant that one attempted to define all theoretical terms occurring in a given theory by means of observational terms.

(2) As a result of a series of unsolvable difficulties, this reductionist stance was later abandoned. Instead, a new approach was suggested which acknowledged the relative independence of theoretical terms from observational ones. Observational terms were thought to refer directly to observable objects in a way which did not need the use of instruments; that is, according to this view, observable properties had to be identifiable without measurement.[2] Examples of such terms are *red*, *warm*, *hard*, *shorter than*, *moving*, etc. Accordingly, theoretical terms were defined negatively and were identified with expressions which could not be related to immediate sensory experience such as *electronic field*, *atom*, *electron*, *mass*, *force*, etc. Consequently, the theory-dependence of theoretical terms was acknowledged.

(3) Nevertheless, it was realized that *all* observation was theory-laden and, therefore, the distinction mentioned in (2) turned out to be untenable. Thus, the next important stage in the development of the problem was necessarily an approach which gave up the dichotomy between observational and theoretical terms completely. The 'non-statement view' (also called 'the structuralist view of theories') developed by J. D. Sneed, W. Stegmüller and others, abandoned the notion of observational terms and defined theoretical terms in the following way:

> The function *n* is theoretical with respect to [a theory, A. K.] *T* if and only if there is no application *i* of *T* in which n_i is *T*-independent; *n* is non-theoretical with respect to T if and only if there is at least one application *i* of *T* in which n_i is *T*-independent. (Sneed 1971: 33)

This means, in a very simplified manner, that a property is to be regarded as theoretical if it is determined solely by the context of a given theory. Unfortunately, this definition is circular: theoretical terms, which constitute a given theory, are determined by this theory itself.

(4) However successful the non-statement view of theories is, this and further difficulties call for revision. A recent approach obtains the following conclusion:

> [...] the problem of theoretical terms is no longer real. When the distinction between theoretical and non-theoretical terms is resolved, this problem vanishes and opens the way to a new understanding of theories. (Zoglauer 1993: 8; my translation, A. K.)

What does this development of the problem of theoretical terms boil down to? By way of summary, let us make the following observations.

- The problem is of a basically *linguistic* nature, because it focuses on the semantics of terms which are linguistic entities.
- As we have seen, in the course of its development the focus of the problem has changed. Whereas early logical empiricism mainly emphasized the referential aspect of theoretical terms, later developments pushed their *context-dependency* into the foreground (see also Percival 2001).
- However, despite its clearly linguistic nature, the problem of theoretical terms was not conceived of as such, but rather as a quandary in the *philosophy of science.*
- Nevertheless, the view quoted above in (4), according to which the problem of theoretical terms seems to lose its relevance for the philosophy of science, may be interpreted in at least two different ways. Firstly, as Zoglauer (1993) suggests, it will be possible to develop philosophies of science which focus on different, and perhaps more substantial problems; evidently, such a task must not be our concern here. Secondly, we can say that it is now fully legitimate to restore the basic linguistic nature of the problem by *reconstructing it as an essentially linguistic issue.*
- Undoubtedly, such a reconstruction will have to raise the general topic of *word meaning.* Moreover, since the role of context has to be acknowledged, the relevance of the relationship between *semantics and pragmatics* must not be excluded at the outset. Finally, this means that what has to be analyzed is a specific type of context, namely, *scientific theories*, and the specific way in which such contexts affect word meaning in the light of the semantics-pragmatics relationship.

Before we proceed, another terminological remark should be made. There is hardly anything more difficult than defining terms like 'semantics' or 'pragmatics' in an a-theoretical or pre-theoretical way. In accordance with our main strategy both terms will be used pre-explicatively, just like other terms already introduced ('cognitive linguistics', 'cognitive semantics', 'science', 'empirical' etc.). Nevertheless, let us, firstly, state the trivial characterization that 'semantics' involves the 'meaning' (whatever 'meaning' is) of expressions and 'pragmatics' is associated with 'intentions', 'goals', 'priorities' expressed by utterances. However unsatisfactory these vague characterizations are, they seem to be sufficient for the moment. All the more so, because, secondly, in Chapters 4 and 5 two very different, theory-dependent characterizations of 'semantics' and 'pragmatics' will be discussed.

3.3 On scientific explanation

Scientific explanation belongs to the most important and at the same time most problematic phenomena discussed in the philosophy of science. The most famous model of scientific explanation which has been widely discussed over the past four decades and which counts as the "point of departure for all discussions of nonstatistical explanation in the philosophy of science" (Newton-Smith 2001b: 127) is Hempel's deductive-nomological (D-N) model (Hempel 1965). The main properties of the D-N model are as follows.

Firstly, scientific explanation is considered to be an inference which has the following structure (for concise overviews see e.g. Bird 1998, Newton-Smith 2001b and for a comprehensive analysis Stegmüller 1983).

Premises:	C1, C2, …Ck
	L1, L2,…L1
Conclusion	E

The first premise corresponds to the so-called 'initial conditions' and consists of the description of particular facts. The second premise comprises the description of general laws. These two premises constitute the *explanans* i.e. that part of the explanation that explains the fact to be explained. The conclusion *E* is the *explanandum* i.e. the description of the fact that has to be explained by the explanans.

Secondly, this argument has to be deductively valid. Thirdly, the descriptions of the initial conditions have to be true. Fourthly, the laws have to be true as well and have to be relevant for the derivation of *E*.

Nevertheless, as further developments have shown, although this model was very fruitful, all of its features mentioned were fiercely disputed and turned out to be untenable. In particular, among other things, the following difficulties arose (for a detailed discussion see Stegmüller 1983 and for a concise summary e.g. Newton-Smith 2001b):[3]

- During the discussions convincing examples were mentioned which showed that there are successful explanations without laws (Scriven 1962).
- The D-N model could not account for the role causation played in explanation (Achinstein 1983 etc.).
- The D-N model could not account for the fact that explanations very often rest on the use of metaphors and analogies (Hesse 1966, 1980).
- It turned out that explanation depends on 'pragmatic' aspects in a relevant way and to a relevant extent (Stegmüller 1983, Kertész 1988).
- There was no avoiding the conclusion that there are many different forms of scientific explanation which may serve different aims and therefore it is hopeless to reduce these to one general formal pattern.

However sketchy and fragmentary this enumeration of certain fundamental difficulties is, hopefully it serves to indicate that the question of what scientific explanation is and how it works is anything but solved (for a detailed and comprehensive overview of the classic discussions up to the eighties see Stegmüller 1983). The present situation can be characterized in a plausible way by quoting Newton-Smith's summary:

> We have an embarassment of riches. We have explanations by reference to causation, to identities, to analogies, to unification, and possibly to other factors. Philosophically we would like to find some deeper theory that explained what it was about each of these apparently diverse forms of explanation that makes them explanatory. This we lack at the moment. (Newton-Smith 2001b: 130–131)

> The prospects of any one of these models being developed to cover all good scientific explanations (let alone all good explanations in general) are dim. (Newton-Smith 2001b: 132)

> The current situation is an *embarrassment for the philosophy of science.* Indeed, one might go so far as to say that it is the sort of *scandal* to philosophy of science that Kant thought skepticism was to epistemology. While we have insightful studies of explanation, we are a very long way from having this single unifying theory of explanation. Why should we want this? As noted above, we would like to be able to explain what it is that leads us to count different explanations as explanatory. This task is made all the more pressing as most philosophers of science hold that a main task, if not the main task, of science is *to provide explanation, whatever that may be. And it is hard to see how we will be able to adjudicate the substantial claims about the relation of explanation to epistemology without such a unifying account.* (Newton-Smith 2001b: 132; emphasis by underlining in the original).

In fact, in spite of the above situation, there is a kind of 'received view' of scientific explanations (Salmon 1997) which is closely related to the problem of theoretical terms and which we may accept as a starting point for our later considerations — while maintaining our reservations.[4] Against this background, we may say that it is the following three properties which the 'received view' of the analytic philosophy of science — independently of particular models of explanation — assumes to be the intuition behind explanations (Salmon 1997): Scientific explanations are considered to be *subsumptive* insofar as they subsume individual facts under generalizations and it is these generalizations which contain theoretical terms; they should at least strive to be *deductive* in that they deduce the explanandum from the explanans;[5] finally, explanations are considered to be structurally identical to predictions, therefore, they must be *predictive.*[6]

The upshot of this situation is as follows. We emphasized that the problem of theoretical terms — which the present part of the book will focus on — is closely connected to scientific explanations. As we have just seen, there is no generally acceptable account of the latter — consequently, there is no reliable starting point for our treatment of theoretical terms. Therefore, the only way out is that to try to select some alleged aspects of explanations (for example the three mentioned in the previous paragraph) which the analytic philosophy of science considered to be relevant in some way and which serve at least historically as part of the background in which the problem of theoretical terms can be understood. Admittedly, this is a very vague and unsatisfactory treatment of the situation, but we have no other choice. No more precise or more reliable starting point seems to present itself, but if we dispense with such a starting point in the light of the complicated nature of scientific explanations, we will never be able to say anything about theoretical terms. Therefore, as a starting point we will refer to the 'received view' mentioned above.

3.4 On explication

Explication is one of the methods of introducing terms into a theory. Explications are assumed to play a systematic role in theories. The terms introduced by the method of explication are relevant components of the *generalizations* which make up the theory. Moreover, explications within a theory are not isolated phenomena, but rather, they constitute a *network* of terms without which the basic claims of the theory cannot be formulated.

The main structural property of an explication is that an imprecise expression — which is called the *explicandum* — is replaced by a precise one, the *explicans*. According to the classic work of Carnap and Stegmüller (1959: 15) explications have to fulfill at least four criteria:

- The explicans must be similar to the explicandum to such an extent that it can replace the latter in each case in which originally the explicandum occurred. Nevertheless, the explicandum and the explicans must not be identical; even considerable differences between them are permitted.
- The rules which govern the use of the explicans have to be given in an 'exact' or 'precise' form so that the explicandum can be integrated into a system of scientific terms.[7]
- The explicans is expected to be 'fruitful'. This means first of all that it should allow the formulation of generalizations.
- The explicans should be as simple as possible.

Just as in the case of explanation, this classic account of explication has been fiercely disputed in the philosophy of science (for surveys see e.g. Hanna 1968, Kertész 1991). Nevertheless, also here there is a kind of 'received view' which is closely related to the problem of theoretical terms and which we may consider to be the starting point for our later argumentation. Namely, the explicandum is *'vague'*, the explicans is *'exact' or 'precise'* (whatever this means), and the relationship between the explicans and the explicandum is basically a *'semantic'* one.

3.5 On the problem of theoretical terms in generative grammar

In Section 3.2. the problem of theoretical terms was formulated on a high level of abstraction. However, in this general form the problem cannot be dealt with: what is needed is its specification with respect to well-defined theories and a well-defined set of theoretical terms whose structure may be analyzed by linguistic methods. For the following two very straightforward reasons it might be useful to choose the theoretical terms of generative linguistics as *paradigmatic examples* and to reduce the problem to the structure of these terms. Firstly, one of the central features of Chomsky's programme was that he identified grammars with theories and assumed that their structure was similar to the structure of theories in natural science. Secondly, during the continuous development of Chomskyan generative linguistics this structure underwent considerable changes, and the investigation of these changes may well be instructive with respect to the structure of theoretical terms in linguistics.

As a starting point, let us show very briefly how Chomsky characterizes the relationship between 'grammar', 'theory' and 'hypothetical constructs' at the outset:[8]

> A grammar of the language *L* is essentially a theory of *L*. Any scientific theory is based on a finite number of observations, and it seeks *to relate the observed phenomena and to predict new phenomena by constructing general laws in terms of hypothetical constructs such as (in physics, for example) 'mass' and 'electron'*. Similarly, a grammar of English is based on a finite corpus of utterances (*observations*), and it will contain certain grammatical *rules (laws)* stated in terms of the particular phonemes, phrases, etc. of English (*hypothetical constructs*). These rules express structural relations among the sentences of the corpus and the indefinite number of sentences generated by the grammar beyond the corpus (*predictions*). Our problem is to develop and clarify the criteria for selecting the correct grammar for each language, that is the correct theory of this language. (Chomsky 1957: 49)

This quotation clearly shows that Chomsky intended to construct a theory which reflects the 'received view' of scientific explanations and theoretical terms.

3.6 Summary

Our point of departure for the considerations of the next two chapters will be the 'received view' outlined above and summirized as follows.

(RV) a. Theoretical terms are introduced into a theory by the method of 'explication'. Thereby
- the explicandum is 'vague',
- the explicans is 'exact', and
- the relationship between the explicans and the explicandum is basically a 'semantic' one.

b. Theoretical terms appear in 'scientific explanations'. Scientific explanations are
- 'subsumptive',
- 'deductive',[9] and
- 'predictive'.

c. 'Pragmatics' (however we define it) is not relevant to the structure of theoretical terms.[10]

Thus the next two chapters will try to explore in what way and to what extent a possible holistic and a possible modular approach to the cognitive science of science can account for the problem of theoretical terms in generative linguistics.

It is important to remark that — as mentioned in the Introduction — we *do not* want to claim that theoretical terms in generative linguistics do indeed work along the lines of the cognitive theory of metaphor or, alternatively, the two-level approach to cognitive semantics. Rather, we proceed *hypothetically*: we set out to explore what the case *would be if* (MHH′) or (MMH′) were acceptable. This is in full accordance with our main problem (Q‴) which focuses on the *prospects and limits* of the two approaches at issue and the nature of the case studies indicated in Section 3 of the Introduction.

In Chapter 4 the prospects of the metascientific application of the cognitive theory of metaphor concerning its capability of solving the problem of theoretical terms in generative linguistics will be exemplified, whereas Chapter 5 will center on that of the two-level model. Although the present part of the book focuses on the *prospects* of the metascientific application of the two approaches at issue, *the case studies also serve to illustrate their limits*. Therefore, we will refer to these limits at certain points of the case studies, but — so as to avoid the overcomplication of our reasoning — they will *not* be discussed *until* Chapter 14, whose main task will be to overview the limits of the two approaches. Finally, Chapter 6 will draw some of the conclusions which the case studies suggest.

CHAPTER 4

Case study: A holistic approach to the problem of theoretical terms

4.1 Introduction

In Section 2.3.2 we introduced the metascientific extension of the main hypothesis of the cognitive theory of metaphor (MHH′) which says that scientific knowledge is metaphorically structured.[1] If we relate this thesis to the problem of theoretical terms in generative linguistics, we may narrow down (Q‴) to the central problem of the present chapter:

$(P)_{ch4}$ How can the following questions be answered, *if* the metascientific extension of the main hypothesis of the cognitive theory of metaphor (MHH′) as an example of holism is accepted:
a. What is the structure of explications in generative linguistics?
b. How is the structure of explications related to the structure of scientific explanations in generative linguistics?
c. To what extent are the answers to (a) and (b) related to semantic and pragmatic factors?

As mentioned in Section 3.6, we will not argue for the tenability of the treatment of theoretical terms on the basis of (MHH′): we just want to see what the case *would be if* (MHH′) were accepted. We will proceed along the lines of the questions asked in $(P)_{ch4}$. Section 4.2 will be devoted to the structure of theoretical terms in generative grammar. Section 4.3 centers on the conseqences of the findings concerning the structure of theoretical terms for the structure of scientific explanations. Section 4.4 will give an account of the role which semantic and pragmatic factors seem to play with respect to the structure of explanations and theoretical terms. Finally, Section 4.5 will summarize the results.

4.2 The structure of theoretical terms

We will analyze some typical theoretical terms of the Standard Theory of generative grammar (Chomsky 1965) and Government-Binding Theory (Chomsky 1981). Thereby we will rely on previous considerations of the same issue in Riley (1987) and Junker (1992). It would be beside the point to introduce our relatively short and simple analyses by summarizing the main assumptions of these two versions of generative grammar — all the more so, because they are quite well-known. Nevertheless, it is useful to highlight at least those of their aspects which may make the outcome of the analyses to follow more understandable. To put it simply, apart from the structure of the grammar and the techniques that are used, on a very abstract level we may say that the major difference between the perspective of the Standard Theory and that of Government-Binding Theory can be characterized in terms of three factors: rule-orientation vs. principle-orientation; the language-particular vs. language-invariant nature of descriptions; the construction-particularity vs. construction-invariancy of descriptions.

With respect to these properties, Chomsky himself compares the two approaches in a very plausible and concise manner:

> We can see more clearly what is at stake by considering two properties that descriptive statements about language might have. They may be language-particular or language-invariant — call this property [±lp]; and they may be construction-particular or construction-invariant — call this the property [±cp]. The quasi-rules of traditional grammar are typically [+lp], [+cp], and *the same is true of the phrase structure rules of early generative grammar*. Thus the rule VP → V-NP is particular to certain languages and a certain construction, namely, phrasal verbs. [...]
> In a principles-and-parameters theory, in contrast, there are *general principles of language that are [−lp] and [−cp]*, and there is the specification of parameters, which is [+lp] and [−cp]. There is nothing else. The only property of descriptive statements is [±lp]. Constructions, in the traditional sense, may be simply an *artefact*, perhaps useful for descriptive taxonomy, but nothing more. If this proves to be correct, traditional grammatical constructions are on a par with such notions as terrestrial animal or large molecule, but are not natural kinds. There is no passive construction, interrogative construction, etc. Rather, the properties of the relevant expressions follow from the interaction of language invariant *principles*, with *parameters* set. The property [±cp] *disappears*. Notice that the property [+lp] cannot disappear. (Chomsky 1991: 23–24; emphasis added)

In analyzing the theoretical terms of the Standard Theory, Riley (1987: 177) lists the terms in (1).

(1) *cycle, daughter, deep, flip-flop, hopping, insertion, inversion, kernel, node, pruning, raising, shift, sister, support, surface, tree, underlying*

Evidently, the cognitive theory of metaphor characterizes these terms as parts of metaphorical expressions which are based on metaphorical concepts. The question, then, is this: Which metaphorical concepts underlie these expressions? The answer is not difficult to find, for a simple reflexion on the terms in (1) may suggest that the target domain is STRUCTURE whereas the source domain is something which is closely connected to the growth, movement or change of living organisms. Thus, we obtain the following metaphorical concepts which are in full accordance with Junker 1992).[2]

(2) Metaphorical concept: STRUCTURES ARE PLANTS
Target domain: STRUCTURES
Source domain: PLANTS
Examples of metaphorical expressions:
The set of <u>kernel</u> S's in a language is the set of S's produced by applying only obligatory transformational rules to underlying structures.
A <u>tree</u> diagram, or phrase marker, is a hierarchical representation of a structure described by either a phrase structure rule or a transformational rule.
A <u>cycle</u> is the domain of application of rules within a structure containing one or more embedded S's.
A <u>node</u> is a labelled point in a tree structure.
<u>Pruning</u> is the elimination of a node which, as the result of a transformational rule, no longer dominates any material.

(3) Metaphorical concept: STRUCTURES ARE HUMAN ACTIVITIES
Target domain: STRUCTURES
Source domain: HUMAN ACTIVITIES
Examples of metaphorical expressions:
<u>Support</u> (insertion) is the introduction of 'dummy' elements (such as auxiliary 'do' or the subject noun phrase 'it') into a structure.
<u>Raising</u> is the movement of an element from an S at a lower level in a tree diagram to an S at a higher level.
<u>Hopping</u> is the movement of an affix from its base-generated position onto the verb that follows it. (Similar terms are *flip-flop, inversion, shift.*)

(4) Metaphorical concept: STRUCTURES ARE FAMILY RELATIONS
Target domain: STRUCTURES
Source domain: FAMILY RELATIONS

Examples of metaphorical expressions:
If node B is immediately dominated by a node A, then B is the daughter of A.
If two or more nodes are daughters of the same node, then they are sisters.

(5) Metaphorical concept: STRUCTURES ARE LEVELS OF HUMAN CONSCIOUSNESS
Target domain: STRUCTURES
Source domain: LEVELS OF HUMAN CONSCIOUSNESS
Examples of metaphorical expressions:
A deep or underlying structure is one generated by the base component, or PS rules, of a grammar.
Surface structure represents the final result of any transformational rules that have been applied to an underlying structure.

These analyses are in full accordance with what Lakoff and Johnson's approach suggests: the metaphorical expressions are manifestations of metaphorical concepts whose source domain is closely connected to our experiences concerning the behaviour of living organisms — that is humans and plants.

Let us compare the corresponding terms of Government-Binding Theory with the observations in (2)–(5). Riley (1987: 177) lists the following.

(6) *barrier, binding, command, condition, constraint, control, filter, government*

These theoretical terms seem to be related to metaphorical concepts which are very different from those in (2)–(5):

(7) Metaphorical concept: STRUCTURE IS SOCIAL HIERARCHY
Target domain: STRUCTURE
Source domain: SOCIAL HIERARCHY
Examples of metaphorical expressions:
B c-commands (constituent-commands) C if the first branching node dominating B dominates C, and B does not dominate C (or vice versa).
Control is the coreference relationship that holds between an empty pronominal NP (PRO) in an infinitive complement and its antecedent.
Governing nodes are Verb, Preposition, Noun, Adjective, Tense and Possessive.
A governs B if A is the minimal governing node c-commanding B.

(8) Metaphorical concept: structure is restriction
Target domain: STRUCTURE
Source domain: RESTRICTION

> Examples of metaphorical expressions:
> *A barrier is an NP or S-bar boundary that blocks government.*
> *An argument (NP position) is bound if it is coindexed with a c-commanding argument; otherwise it is free.*
> *Constraints/conditions are used to to describe structures which prohibit the application of a rule during derivation.*
> *A filter is a restriction on the surface structure resulting from derivation.*

It is important to emphasize that the instances of scientific concept formation are not to be confined to individual metaphors but, rather, they are to be identified with complex conceptual networks including different cognitive models (cf. (18) and (20c) in Section 2.3.1). In particular, (20) in Section 2.3.1 yields the assumption that it is not only the theoretical terms themselves which matter, but also the system of metaphorical expressions in which they occur. Thus the theoretical terms listed in (1) and (6) in the present chapter are manifestations of metaphorical concepts whose main task is the conceptualization of the subject matter of linguistic theory, namely, grammar. Nevertheless, as we have seen, the theoretical terms of the Standard Theory and Government-Binding Theory conceptualize 'grammar' in two very different ways. Theoretical terms in the Standard Theory are rooted in metaphorical concepts which conceptualize grammar with respect to *creativity, dynamism, motion, growth* (cf. also Riley 1987). Theoretical terms in Government-Binding Theory are based on metaphorical concepts which conceptualize grammar as being something which is exposed to *limitations, restrictions, confinements* (cf. also Riley 1987, Junker 1992).[3]

We have noticed in Section 3.4 that the general method of introducing theoretical terms into a theory is that of *explication.* How can the process of explication be interpreted if the theoretical terms of generative grammar are conceived of as the manifestations of metaphorical concepts? In order to find an answer, let us quote Riley, again:

> [...] Chomsky and other linguists have adopted the jargon of transformational syntax from other vocabularies, and in doing so *have endowed each term with a specialized meaning, one not necessarily meant to convey the denotative or connotative properties of the term as it is used outside of linguistics.* (Riley 1987: 184; emphasis added)

If we reinterpret this quotation with respect to the cognitive theory of metaphor, we can make the following observations.

The picture we get is entirely different from the assumptions of the analytic philosophy of science. Though the explicans is to be identified with the 'technical', i. e. theory-dependent, use of the term in metaphorical expressions, the point is not that the explicans is 'precise' or 'exact' (as the 'received view' of the analytic philosophy of science would maintain). Accordingly, in the light of the cognitive theory of metaphor the relevant feature of the explicandum — the original, nontechnical use of expressions like *deep*, *govern*, *command* etc. — is not the fact that their 'vagueness' is to be contrasted with the 'exactness' or 'precision' of their theoretical use. Moreover, the relation between these two terms — i.e. the explicans and the explicandum in the sense of the analytic philosophy of science — is not considered to be relevant in the light of the cognitive theory of metaphor. The relevant relation holds between the theoretical term and the metaphorical concept underlying it. What matters, is that terms are introduced into the theory which are manifestations of particular metaphorical concepts in metaphorical expressions (in the sense of (20a) in Section 2.3.1). However this relation is conceived of, it is assumed to be 'holistic' and therefore it differs from the 'semantic' relationship presupposed by the 'received view' (RV).[4]

Thus we obtain a possible answer to $(P)_{ch4}$(a) which says that *if* the cognitive theory of metaphor were accepted as a possible approach to the cognitive science of science, then the relevant aspect of theoretical terms would not be the relation between a 'precise' / 'exact' term and a 'vague' one, but that between a metaphorical expression and a metaphorical concept.

4.3 The structure of explanations

The answer to $(P)_{ch4}$(b) may be inferred from theses (23) and (25) in Section 2.3.1 immediately: in scientific explanations it is the target domain *X* of a metaphor which is to be regarded as the explanandum whereas the source domain *Y* plays the role of the explanans:

> As a rule, the metaphor (*X is Y*) relates an abstract and complex target domain (*X*) as an *explanandum* to a concrete source domain with a simple structure and exposed to our sensory experience (*Y*) as an *explanans*. Thereby the relation between the elements *X* and *Y* is irreversible, the metaphorical projection is unidirectional. (Jäkel 1997: 57; emphasis by underlining original; my translation, A. K.).

This amounts to the assumption that metaphors provide us with image schemas which help us during linguistic research to make an unknown phenomenon *X* understandable, insofar as it is related to a very different domain *Y* which can be experienced immediately or which is, at least in a certain respect, more familiar than *X*. This is what seems to happen, for example, with respect to source domains like PLANTS, HUMAN ACTIVITIES, HUMAN ROLES, etc. in the Standard Theory.

An important consequence of this is the way in which metaphorical concepts may fulfill their focussing function (see (27) in Section 2.3.1). Clearly, the properties mentioned are rendered salient in the explanations, too: in the Standard Theory, the creative, dynamic nature of grammatical structures is emphasized, whereas the explanations of Government-Binding Theory focus on their restrictiveness.

If the metascientific extension of the main hypothesis of the cognitive theory of metaphor were accepted, then we would obtain the following answer to $(P)_{ch4}$ (b):

What the cognitive theory of metaphor considers to be relevant with respect to explanations is the mapping between the explanans and the explanandum — which are conceptual domains. Thus if one accepted (MHH′), then an entirely different picture of the nature of explanations would be drawn than that assumed by the "received view". Nevertheless, since in the light of (27) in Section 2.3.1 metaphorical concepts focus on just one specific aspect of the target domain, which is identified with the explanandum of an explanation, one may predict that all future occurrences of the phenomena thus conceptualized will have the very same aspect. Therefore, grammatical explanations — seen as metaphorical concepts along the lines of (23) and (25) — appear to be, in this sense, predictive.

4.4 Theoretical terms and pragmatics

By definition, holism denies the existence of boundaries between what might be called 'semantic', 'cognitive' or 'pragmatic' aspects of the knowledge of language. For example, Lakoff and Johnson characterize their approach in the following way.

> It [i.e. their approach, A. K.] includes, for starters, all those things you would have to learn if you were to learn a foreign language: *the meanings, the pragmatics, the speech-act constructions, constraints on processing, and on and on.* (Lakoff and Johnson 1999: 482; emphasis added)

Accordingly, if, on the basis of (MHH′), the theoretical terms of generative linguistics are described by means of a holistic approach to the 'cognitive science of science', then we have to capture the 'wholeness' of at least (a) the metaphorical expressions which constitute the system of theoretical terms of generative linguistics, (b) the metaphorical concepts governing them, and (c) the scenarios which are to be regarded as the overall cognitive framework with respect to which metaphorical concepts and metaphorical expressions work.

No matter how we define the notion of pragmatics, it goes without saying that 'goals' and 'priorities' belong to pragmatics. Now, as Riley shows, the terminology of both the Standard Theory and Government-Binding Theory reflect the goals of linguistic theory in a clear-cut way. As regards the Standard Theory, Riley emphasizes "Chomsky's interest in what could properly be called the *creative* or *dynamic* properties of language systems" (Riley 1987: 181). The result of comparing the terminology of the Standard Theory with the goal of capturing the creative and dynamic nature of the knowledge of language boils down to something which Riley expresses in the following way.

> Regarding language as an intricate, creative, yet rule-governed system, linguists within the ST [the Standard Theory, A. K.] framework have adopted and introduced a metalanguage that is highly compatible with their paradigm. The recurring qualities of ST jargon are those associated with the properties that distinguish *living organisms* from nonorganic entities: *process, movement, and growth. The dynamic and creative qualities of ST jargon thus strongly parallel the central concerns of the theory itself, its emphasis on the creative aspects of language.* (Riley 1987: 181; emphasis added, A. K.)

This insight may be used to reveal the general scenario (in the sense of (28) in Section 2.3.1) which includes at least the following components:

- INITIAL STATE: there are living organisms
- PATH: these living organisms move, grow, develop
- FINAL STATE: the result of this growth, development and creativity, i.e. richness and complexity

We may adapt Riley's findings to Government-Binding Theory in an analogous way. Accordingly, the main goal of Government-Binding Theory is the restriction of grammar. It is instructive to notice how Chomsky himself contrasts the *goal* of the Standard Theory with that of Government-Binding Theory:

> The theory presented in [the *Logical Structure of Linguistic Theory* (1955)] permitted a great number of rules. I tried at first to provide a system rich enough to express as much as I could imagine. Now, in a sense, I'm trying to do the opposite, *to limit the expressive power of rules* (Chomsky 1979: 182; quoted in Riley 1987: 182; emphasis added, A. K.).

The scenario:
- INITIAL STATE: there are abstract social relations,
- PATH: these social relations are restricted,
- FINAL STATE: impoverished system of relations.

The 'pragmatic' aspects of theoretical terms are rooted in the very nature of metaphorical concepts and the scenarios associated with them. Thus the answer to $(P)_{ch4}$ (c) is that the cognitive theory of metaphor does not differentiate strictly between the semantic, conceptual and pragmatic aspects of theoretical terms and explanations. Consequently, pragmatics is one of the factors which play a constitutive role in scientific concept formation.

4.5 Summary

The case study carried out in the previous sections led to the following solution to $(P)_{ch4}$:

$(SP)_{ch4}$ If the metascientific extension of the main hypothesis of the cognitive theory of metaphor (MHH′) is accepted, then in generative linguistics

a. the relevant aspect of theoretical terms is that they are assumed to be metaphorical expressions which are manifestations of metaphorical concepts;
b. in accordance with (a) the explanans is identified with the source domain, the explanandum with the target domain, and there is a mapping between these two domains;
c. in accordance with (a) and (b) the cognitive theory of metaphor does not differentiate strictly between the semantic, the conceptual and the pragmatic aspects of theoretical terms and explanations, and it acknowledges, therefore, in this 'holistic' sense, the role of pragmatics in scientific concept formation.

Although the case study in the present chapter centers on the prospects of the metascientific application of the cognitive theory of metaphor, these prospects cannot be evaluated without realizing at least some of the limits of the analyses presented. Nevertheless, it would be beside the point to survey the general criticism of the cognitive theory of metaphor put forward in the literature. Instead, in accordance with what has been said in the *Introduction*, our task is to *illustrate* the nature of the limits of the metascientific application of the cognitive theory of metaphor which *the above case study itself* illuminates. However, as already mentioned, we will postpone the discussion of these limits to Chapter 14.

Chapter 5

Case study: A modular appoach to the problem of theoretical terms

5.1 Introduction

In this chapter we proceed analogously to Chapter 4. First of all we *assume* that the metascientific extension of the generalized modularity hypothesis (MMH′) introduced in Section 2.2.2 can be accepted. Then we have to explore the answers which under this assumption can be given to the three questions asked in $(P)_{ch5}$.[1]

$(P)_{ch5}$ How can the following questions be answered, *if* the metascientific extension of the generalized modularity hypothesis (MMH′) as an example of modularism is accepted:

a. What is the structure of explications in generative linguistics?
b. How is the structure of explications related to the structure of scientific explanations in generative linguistics?
c. To what extent are the answers to (a) and (b) related to semantic and pragmatic factors?

Also our line of argumentation will be analogous to that in Chapter 4: the next three subsections will deal with the three questions in $(P)_{ch5}$, respectively, and the last section will summarize the solution to $(P)_{ch5}$. Nevertheless, due to the nature of the theoretical framework, the internal structure of Sections 5.2 and 5.3 will be somewhat more complex than that of the corresponding sections in Chapter 4.

5.2 The structure of theoretical terms

5.2.1 The semantic underdetermination of theoretical terms

One corollary of (MH′) is the thesis which we mentioned in (11)(i) in Section 2.2.1. This assumption says that the interpretation of a lexical item *A* of language *L* in context *ct* is the mapping of its semantic representation *sem* onto the level of conceptual structure *CS* in such a way that the semantic representation *sem* yields the conceptual representation *m* of a lexical item *A* of language *L* relative to context *ct*, that is: *sem(A, ct)=m*.

According to the metascientific extension of the generalized modularity hypothesis (MMH′) (see Section 2.2.2) this corollary must also apply to the language of scientific theories. Therefore, we should adopt it with the only remark being that *L* is the language of an objectscientific theory — say, the Standard Theory or Government-Binding Theory of generative linguistics — and *A* is a term of this language. We have also seen in (11) in Section 2.2.1 that what may be called the 'meaning' of a term is semantically underdetermined, because it results from the interaction of the grammatical and the conceptual module. This semantic underdetermination leads to far-reaching consequences with respect to the nature of theoretical terms.

So as to understand these consequences, let us remember some basic assumptions of the two-level approach. Firstly, it is assumed that the semantic representation *sem* of a term *A* determines a family of conceptual representations (a family of concepts, see (12) in Section 2.2.1) — which are different context-dependent interpretations of *A*. Secondly, the relations between these conceptual representations are governed by principles called conceptual shift, conceptual specification and conceptual selection.

Now the question is how, against this background, the metascientific extension of the two-level approach (MMH′) answers $(P)_{ch5}(a)$. Naturally, the answer has to be restricted to theoretical terms whose 'meaning' belongs to those phenomena which the two-level approach focuses on. Then, in such cases, the relationship between two different explicantia of a given theoretical term *A* may correspond to that between two conceptual representations of this term which are members of the same family of concepts *FC* determined by a common semantic representation *sem*. For example, two explicantia may be connected to each other by the principle of conceptual shift: that is, the *sem* which they share is mapped onto two different theoretical contexts and as a result of this, the interpretation of the term *A* is shifted to two different 'conceptual domains'. Such a term does not seem to be 'exact' in the sense of the analytic philosophy of science, because it can be interpreted in many different ways.[2] However, there may be cases in which it is neither ambiguous between two meanings inherent in its lexical structure nor vague, because there is a difference between the two 'meanings'; rather, it is simply assigned different interpretations according to the 'conceptual domains' it is mapped onto.[3] Thus, we may conclude that in the light of the metascientific extension of the generalized modularity hypothesis (MMH′), *the structure of explications* of a given theoretical term is characterized by two aspects. Firstly, as opposed to the claims of the analytic philosophy of science, what has traditionally been labelled the *explicandum*, is assumed to be neither something vague nor something ambiguous. Rather, it can be identified with the semantic representation *sem*. Secondly, in such cases the *explicans*, as contrasted with the 'received view' (see Section 3.4), is not an 'exact' term, but one of the conceptual representations yielded by one or another of the conceptual operations mentioned above.[4] That is, we obtain the result that the explicandum of a term *A* is its *sem* and its different explicantia are its conceptual representations m_i — *which, nevertheless, belong to the same family of concepts.* From these findings it follows that the metascientific application of the two-level approach yields an answer to $(P)_{ch5}(a)$ which says, firstly, that theoretical terms in generative linguistics are *not* introduced by simply replacing a 'vague' term of everyday language by an exact one. The second consequence is that the relationship between the explicans and the explicandum is *not* semantic in the sense of the 'received view'; rather, it is semantically underdetermined, because it rests on the relationship between the semantic and the conceptual representation.

The analytic philosophy of science cannot explain how it is that, on the one hand, the 'exactness' of terms is a major requirement, while on the other hand, theories work quite well even if the terms they use are not 'exact'. The lack of exactness would be fatal only if in a given context the interpretation of the term were not clear and if there were no connection between the various interpretations. But if we try to understand the structure of theoretical terms with respect to families of concepts as we have done above, neither case will apply. Thus we obtain the following.

- Theoretical terms may be assigned different conceptual representations according to the contexts in which they are used. These interpretations are related to each other by conceptual principles such as the principle of conceptual shift, conceptual specification and conceptual selection.
- The conceptual principles do not relate the particular conceptual representations directly, but only indirectly, through *sem*.
- Consequently, the different conceptual representations of a theoretical term *A* are not independent of each other and thus their continuity is accounted for.

So as to illuminate the plausibility of the above argumentation, we have to exemplify how the mechanism outlined works. In the following examples many details are not elaborated on and the analyses draw a simplified picture of the structures we have to reveal. Nevertheless, these difficulties are inherent in the two-level approach which applies a kind of *quasi-formal* method of analysis; thereby certain tools of formal semantics and componential analysis are adopted, but those details which are not focussed on with respect to the particular problem raised are left open. Since our task is to examine what would follow with respect to the structure of theoretical terms if the two-level approach were accepted, we simply have to stick to the points of departure which have been suggested in the literature and proceed analogously; it is not our task to go into the details which the proponents of the two-level approach themselves did not specify or to answer the questions which they left open.

5.2.2 An example: The family *command*

Basically, the assumptions summarized in (16) in Section 2.2.1 suggest that there may be a specific relationship between metaphorical and literal meaning with respect to theoretical terms. Firstly, within the frames of the two-level approach it follows from what has been said that their 'metaphorical meaning' results from the 'literal meaning' of the everyday use of the corresponding terms. Secondly, however, in the second quotation from Bierwisch (1983a: 67) in Section 2.1, central theoretical terms in linguistics which at first sight would seem to be metaphorical *are assumed to work as if they were used literally* and they are, therefore, characterized by being semantically underdetermined; see, for example, *competence* or *performance.* This means that after they have been introduced into a theory by metaphorization, such terms may be interpreted as 'literal' expressions relative to the given theoretical context (i.e. the notions, theses, background assumptions, data etc. which make up the theory at issue) and therefore may be exposed to the sequential application of the operations of conceptual specification and conceptual shift. Thirdly, Bierwisch (1979: 143) explicitly emphasizes that the interpretation of expressions relative to a non-neutral context does not follow linearly from its interpretation with respect to a neutral context in each case, but sometimes there may be much more complex relations between literal and non-literal interpretations (see also (16) in Section 2.2.1). Although the principles conceptual shift, conceptual specification and conceptual selection operate only on literal meaning, it must not be excluded at the outset that they may be applied to expressions which involve a metaphorical process as well, but which in certain contexts are treated as if they were literal. Fourthly, the background assumptions of the two-level approach and its emphasis on semantic underdetermination seem to suggest that the two-level approach prefers to explain those central properties of theoretical terms which are responsible for their structure and functioning in theoretical contexts by attributing relevance to their semantic underdetermination rather than to their metaphorical origin. Fifthly, an independent argument may be mentioned by quoting Tsiapera (1990). Tsiapera showed, among other things, that while during the early development of historical linguistics, organic metaphors played a substantial role in theory formation, later on in the Neogrammarian period the theoretical terms lost their metaphorical nature. From this she concludes that at a certain point in the development of theories metaphors are abandoned:

> In the early period of historical linguistics, the organic metaphor was useful in the initial construction of linguistic theory. Once the comparative method reached a certain level of refinement through verification on empirical grounds, it outgrew the boundaries of imprecision by the metaphor and the metaphor was abandoned. We may ask ourselves, what are the boundaries which have delimited or delimit the usefulness of [...] metaphorical concepts in linguistic theory, for example the 'waves' of wave theory, or the 'depth' of deep structure? To the extent that theory is inevitably expressed verbally, and that natural language abounds in metaphor, metaphors will continue to play an essential part in the development of our science. And yet we may argue that a true advance in science, including in language study, is reached at the point where the method developed by a theory attains a degree of precision which renders the metaphor obsolete. (Tsiapera 1990: 585–586).[5]

In this respect the history of generative linguistics may be analogous to the development of historical-comparative linguistics, because it may be the case that its theoretical terms abandoned their metaphorical nature.

It is worth noting that this assumption is not compatible with the cognitive theory of metaphor which claims that "[...] the concepts in a scientific theory, are often — *perhaps always* — *based on metaphors* [...]" (Lakoff and Johnson 1980a: 19; emphasis added). In contrast, within the framework of the two-level approach it seems to be well-motivated to investigate in what way and to what extent theoretical terms can be analyzed as being *based on semantic underdetermination.*

In the light of these considerations, let us turn to an instructive example, namely, the theoretical term *command.* This example serves to illustrate how a series of conceptual operations may interact and how as a result of this interaction a complex family of conceptual representations is constructed. We will see that each element of such a family may be regarded as a specific explicans of the explicandum, that is, the semantic representation *sem* of the term at issue. We will start from the assumption that the term *command* is treated in the theoretical context as if it were literal and therefore, it is exposed to the consequences of that kind of semantic underdetermination that has been illustrated by the quotations in Section 2.1 and explained in (12)–(15) in Section 2.2.1.

In different stages of the development of generative grammar, among other things, the following interpretations of this term were proposed which we quote from the works referred to.[6]

(1) We shall describe the scope of wh- and *neg* in terms of the concept 'in construction with' [...]. A constituent [...] is 'in construction with' another constituent [...] if the former is dominated by (that is, occurs somewhere lower down the branch of) the first branching node [...] that dominates the latter [...]. (Klima 1964: 279)

(2) We will say that a node *A* 'commands' another node *B* if (1) neither *A* nor *B* dominates the other; and (2) the S-node that most immediately dominates *A* also dominates *B*. (Langacker 1969: 167)

(3) (70) *John knows what who saw.
(71) John remembers where Bill bought which book.
(72) John remembers to whom Bill gave which book.
[...] we say that the category *A* is 'superior' to the category *B* in the phrase marker if every major category dominating *A* dominates *B* as well but not conversely. (Chomsky (1973: 246)

(4) (13) A node *A* is *superior to* a node *B* if the first branching node dominating *A* also dominates B. (Reinhart 1974: 94)

The definition in (13) is identical to the definition of the relation *in construction with* which was suggested by Klima (1964). Chomsky's definition of *superiority* (1973) is distinct in excluding sister nodes from the definition of superiority. But since at least in one coreference case — that of topicalization — that rule which is proposed here applies to sister nodes, I use Klima's and not Chomsky's definition. An alternative [...] name for the relation defined in (13) is C(onstituent)-command, which was suggested to me by Nick Clements. (Reinhart 1974: 105).

(5) A constituent *A* is in construction with a constituent *B* if *A* is dominated by the first branching node which dominates *B*, and *B* does not dominate *A*. (Wexler, Culicover and Hamburger 1975: 243).

(6) *A* commands *B* if the minimal cyclic node dominating *A* also dominates *B*. (Lasnik 1976: 15; here NP and S count as 'cyclic').

(7) A node *A* c(onstituent)-commands a node *B* iff the first branching node a that dominates *A* either dominates *B*, or is immediately dominated by a

node a¢ which dominates *B*, where a and a¢ are of the same category type. (e.g. S and S¢). (Reinhart 1983: 50)

(8) *b* is said to *c-command a if P* does not contain a [...*]and a* is dominated by the first branching category dominating *B.* (Chomsky 1980a: 10).

(9) *x* c-commands *y* iff *(z)*, *z* is a maximal projection, *z* dominates *x* only if it dominates *y*, and *x* ffi *y*. (Aoun and Sportiche 1982: 224)

(10) *a c-commands b* if and only if
i. *a* does not contain *b*
ii. Suppose that $g_1,\ldots g_n$ is the maximal sequence such that
a. $g_n = a$
b. $g_i = a^j$
c. g_i immediately dominates g_{i+1}.
Then if *d* dominates *a*, then either (I) *d* dominates *b* or (II) $d = y_i$ and y_i dominates *b*. (Chomsky 1981: 166)

(11) *a* c-commands $b =_{\mathrm{df}}$ every maximal projection dominating *a* dominates *b* and *a* does not dominate *b*. (May 1985: 34).
[...] to be dominated by an occurrence of a projection, maximal or otherwise, is to be dominated by all the member nodes of that projection. [...] Now, assuming that *a* and *b* are maximal projections, the c-command domain (and hence the scope, under appropriate circumstances) of *B* is the *b*-projection; it therefore includes *A*, *C*, and *D*, which are also dominated by this projection. The c-command domain of *C* is the *a*-projection, since it is dominated by both a^i and a^j. The latter domain includes *D*, but not *B*, which is not dominated by the *a*-projection but by only one of its member nodes, namely *a*. (May 1985: 57)

(12) The domain of *a* is the least maximal projection containing *a*. ...We say that *a* c-commands every element of its domain that is not contained within *a*. (Chomsky 1986: 162)

(13) [...] for two theta roles *X* and *Y*, *X* th-commands *Y* if *X* is a coargument of *Y*; or, if *X* th-commands *B*, *B* is assigned to Z, and *Z* th-commands *Y*. (Williams 1987: 166)

The first observation which we have to make is that there is an important difference between (13) on the one hand and (1)–(12) on the other hand.[7] However different (1)–(12) may be in other respects, all of them refer to *structural relations between constituents.*[8] In contrast, in (13) this relation holds between theta-roles; moreover, in Williams (1987) the term *th-command* was related to *c-command.* Accordingly, it seems to be well-motivated to assume that here we are dealing with the principle of conceptual shift in that the semantic representation *sem* of the term at issue is mapped onto two different conceptual domains in a context-dependent way: one domain is, loosely speaking, the theory of constituent structure (however it may be defined) and the other is theta-theory. Now, let V' be a set of functions which relates the semantic representation of the term *command* to a family of concepts in a way specified in (12) in Section 2.2.1. Then V' includes at least the elements V_{cs} and V_{tht} such that

(14) a. $V_{cs}(sem_c, ct_{cs}) = CR_{cs}$
b. $V_{tht}(sem_c, ct_{tht}) = CR_{tht}$

Here sem_c is the semantic representation of *command*, CR_{tht} stands for the conceptual representation exemplified in (13) whereas CR_{cs} for the conceptual representations of (1)–(12); ct_{tht} and ct_{cs} are the corresponding contexts.

As we have seen in Section 2.2.1, conceptual principles such as conceptual shift and conceptual specification can be combined with each other, and such combinations lead to complex networks of conceptual principles and the conceptual representations of theoretical terms. Accordingly, we may raise the question whether the differences within (1)–(12) can be accounted for by assuming that further conceptual principles operate. In fact, the meanings of *command* in (1)–(12) go back to two traditions: to the so-called Klima-tradition which is 'geometrical' in nature and the so-called Langacker-tradition which combines the geometrical aspect with that of constituent type. Thus within (1)–(12) conceptual shift is applied again in that the semantic representation is mapped onto two different conceptual domains. It seems to be the case that the definitions (1), (3), (4), (5), (8) are 'geometric' whereas (2), (6), (9), (10), (11), (12) focus both on 'geometric' relations and 'constituent type'. How can we differentiate between the conceptual representations within these two subfamilies? In (14) in Section 2.2.1 we said that whereas conceptual shift maps the semantic representation of a term on different 'conceptual domains' i.e. different 'categories' or 'types of concept', conceptual specification differentiates between conceptual representations within the same domain or category type of concepts and 'fills in open slots'. In the light of this it seems to be plausible to assume that within the two conceptual domains mentioned we don't find another application of conceptual shift; rather, the conceptual representations within the subfamilies 'geometrical' and 'geometrical + constituent type', respectively, seem to be differentiated by the principle of *conceptual specification.* For example, (1) and (3) belong to the same 'conceptual domain' (which we characterized as 'geometrical'), although they are associated with different conceptual representations. The same applies to members of the subfamily of concepts 'geometrical + constituent type'.

Consequently, the metascientific application of the two-level aproach suggests that the internal relations which hold between the members of the family of concepts associated with the theoretical term *command* may be obtained through capturing the way principles like conceptual shift and conceptual specification work. (Some of the possible counter-arguments against such an analysis will be mentioned in Section 14.3)

5.2.3 On the degree of semantic underdetermination

So as to avoid the oversimplification and overgeneralization of our results, they have to be interpreted in a more sophisticated way. In particular, the assumption that in generative grammar theoretical terms are semantically underdetermined must not mean that the degree of semantic underdetermination is the same in the case of all theoretical terms (see (11) and (15) in Section 2.2.1 on the notion of semantic underdetermination). Obviously, there are great differences between lexical units with respect to their ability to allow themselves to be conceptually shifted or specified (see Bierwisch 1983b: 92). Therefore, we have to deal with a wide range of possible degrees of semantic underdetermination.

In the theoretical language of generative grammar at one end of the range we find cases such as *noun phrase* which do not seem to allow the variaton of interpretation in a significant way. At the other end there are theoretical terms like *command* which may be associated with a rich family of conceptual representations related to each other by a complicated network of conceptual operations. For example, *sister* or *mother* or *daughter* etc. may not be underdetermined at all, or are at least not as underdetermined as *command*, *bind* or *govern*. However, the assumption that certain theoretical terms are semantically underdetermined to a small and perhaps insignificant degree only, must not imply that we have to exclude them from the phenomena whose meaning can be explained by referring to their semantic underdetermination. If we accept this, we have to ask the question of the relationship between theoretical terms which are strongly underdetermined and those whose semantic underdetermination is relatively weak or insignificant. This question can be subdivided into two more specific problems:

(15) i. Is the degree of the semantic underdetermination of theoretical terms within a given theory invariant, or does it rather change in the course of the development of the theory?

ii. How can the relationship between the scientific and the everyday use of the terms be captured if we know that there are different degrees of semantic underdetermination? More precisely, is the degree of semantic underdetermination of a term greater or smaller in a scientific theory such as (one version of) generative grammar than in its everyday use?

As regards (15i), let us remark first of all in general that the problem of theory change is one of the central and most deeply discussed quandaries of the philosophy of science which we cannot even touch on here. Secondly, in particular, the problem of theory change in generative grammar has been widely discussed as well — here it would be beside the point to overview the literature.[9] We confine the answer to (15i) to mentioning just a few immediate consequences of what we have said about the structure of theoretical terms in generative linguistics so far. We may answer (15i) by saying that the degree of the semantic underdetermination of theoretical terms is historically changing. It is relatively easy to observe that in different stages of the development of the theory the terms at issue had different structures. Therefore, it is plausible to assume that it is the particular degree of underdetermination that contributes to finding the very generalizations which are expected to solve the problems raised in a particular version of generative grammar, or more specifically, explain the phenomena which are focussed on. There are problems which in the theoretical context of a certain stage of development can be solved by introducing theoretical terms with a low degree of semantic underdetermination. Alternatively, in another theoretical context they may be useful if they are treated as maximally underdetermined.

Let us refer, as a plausible illustration of this, to the theoretical term *sentence*. In the early phase of generative grammar *sentence* was explicated within a formal system in that it was considered to be a terminal string which was derivable from an initial symbol *S* by the use of re-writing rules:

> We call *V* a terminal string if there is an *#S#*-derivation of *#V#*, where *S* is the designated *initial symbol* of the grammar (representing the category 'Sentence'), and # is the *boundary symbol* (regarded as a grammatical formative). Thus we construct a derivation of a *terminal string* by successively applying the rewriting rules of the grammar, beginning with the string #S#, until the final string of the derivation consists only of formatives and therefore no further rewriting is possible. (Chomsky 1965: 66; emphases in the original)

Obviously, in such a system the term *sentence* is minimally underdetermined and its meaning is not exposed to variation. However, in Government-Binding

Theory nobody would strive to restrict the meaning of *sentence* to what can be defined within a formal system; rather, it has many different interpretations which depend on the particular contexts in which it is used and therefore it is a typical example of how a theoretical term can be associated with a rich family of conceptual representations (see Stechow and Sternefeld 1988: 100–104 for an instructive discussion of this.). The motivation for accepting the essential polysemy of the term *sentence* which we interpreted here as the result of a high degree of semantic underdetermination was characterized by Stechow and Sternefeld in this way.

> [...] It should be also clear why one does not strive for a formal system: the latter would according to the present development of research dispense with any empirical justification. It is also evident that the lack of a formal system is not identical with the lack of clarity. On the contrary, if clarity is related to intelligibility, then the lack of a formal system is a desideratum, because the latter is, as a rule, quite unintelligible. (Stechow and Sternefeld 1988: 104; my translation, A. K.)

Thus, firstly, it seems to be the case that the degree of the semantic underdetermination of theoretical terms may change in the course of the development of the theory. Secondly, it is not the case that one of the reasons why one of the later phases of a theory may be considered as being more successful than an earlier one is that it makes use of more 'exact' i.e. semantically less underdetermined terms. Thirdly, the crucial question — which we cannot answer here — is in what way and to what extent certain degrees of semantic underdetermination may serve the solution of the problems at issue and, in connection with this, the explanation of the phenomena they focus on. Evidently, there is no mechanical answer to this question: in certain cases a theoretical term may be fruitful and effective, because it shows a high degree of semantic underdetermination, and in other cases, because its capability of being associated with a rich family of conceptual representations is restricted.

As regards (15ii), we have to remember that the two-level approach assumes that both the structure of everyday expressions and that of theoretical terms rest on the same mechanism and do not differ essentially (see the quotations in Section 2.1). From this it follows that, at the outset, it is not possible to assume that theoretical terms are usually semantically less underdetermined than everyday expressions or vice versa. Rather, the particular differences between everyday and expert expressions will change from case to case. For example, the theoretical term *sentence* in the Standard Theory is surely less underdetermined than the same term in everyday discourse. However, as regards Government-Binding Theory, the opposite is true. We may even assume that *sentence* in its everyday use means something like 'sequence of words', whereas in Government-Binding Theory there is a rich variety of its interpretations. Thus, firstly, the metascientific extension of the generalized modularity hypothesis (MMH′) does not allow us to draw mechanically a dividing line between the nature of scientific and non-scientific concept formation and this stance can be easily justified by many plausible examples. Secondly, in connection with this, (MMH′) is incompatible with the traditional view according to which theoretical terms are necessarily more 'precise' or more 'exact' than the everyday use of expressions.

As a result, the metascientific extension of the two-level approach seems to make the structure of theoretical terms appear to be significantly different from what is assumed by the analytic philosophy of science.

5.3 The structure of explanations

Turning to *explanations*, we can claim that the most important function of theoretical terms is that they are part of the 'explanans' of scientific explanations. What matters now is that the main reason why theoretical terms are introduced into a theory is that without them the explanation of a fact described by the corresponding explanandum would not be possible, because theoretical terms enter the explanans in a significant way. What should interest us now is the effect of the above account of theoretical terms with respect to the role they play in scientific explanations. Our considerations will be informal and they will focus on a plausible illustration of the main points rather than on a precise presentation of the details.

Let us assume that Em_1 is the explanandum of the explanation E in which a certain term T_1 occurs and that T_1 has the conceptual representation CR_1. Em_1 should be explained by an explanans Es_1 which contains the theoretical term t_1 with the conceptual representation cr_1. However, both CR_1 and cr_1 are members of families of concepts associated with T_1 and t_1, respectively, in the sense mentioned in the previous section. These families consist of a set of conceptual representations related to each other by conceptual principles such as conceptual shift and conceptual specification. Given the fact that Em_1 and Es_1 are, within the explanation, connected with each other, T_1 interacts with t_1. Consequently, the principle of conceptual selection will operate and according to this principle a given conceptual representation of one term selects a conceptual representation from the family associated with another term. Therefore, it may be the case that cr_1, instead of selecting CR_1, selects another element CR_2 from the family of concepts associated with T_1, while CR_1 and CR_2 are related to each other by one of the conceptual principles. This means that, if this is the case, Es_1 does not explain Em_1 but Em_2, which differs from Em_1 at least insofar as the interpretation of T_1 is not CR_1 but CR_2. The effect of this mechanism is that *an explanans may explain an explanandum not identical with* the one which originally motivated the search for an explanation. Let us illustrate this mechanism by an example.

The theory of binding focuses on the question of why *X* and *Y* are coreferential in (16a) and not in (16b):

(16) a. *Mary$_i$ admires herself$_i$*
X *Y*
b. **Mary$_i$ admires her$_i$*.
X *Y*

The explanandum Em_1 is as follows:

(17) a. In the sentence (16a) *X* and *Y* are coreferential.
b. In the sentence (16b) *X* and *Y* are not coreferential.

In (16) and in (17) *X* and *Y* are syntactic expressions in a structural configuration and *coreferential* is the term T_1 which occurs significantly in Em_1. Without going into an analysis of the historical development of suggestions put forward to explain (17), let it be sufficient to refer to Chomsky's classic solution (Chomsky 1981):

(18) a. An anaphor is bound in…
b. A pronoun is free in …
c. An *R*-expression is free.

Thus the explanation is based on the introduction of the theoretical term *bind* — which we will label as t_1 — into the explanans. The interpretation of t_1 corresponds, on the basis of the metascientific extension of the generalized modularity hypothesis (MMH′), to its context-dependent conceptual representation cr_1 which may be characterized as follows.

(19) *X* binds *Y* iff
i. *X* c-commands *Y*
and
ii. *X* and *Y* are coindexed.

In (19) *c-command* is a structural relation between two nodes in the sense of Chomsky (1981). Accordingly, we get the explanans Es_1 which yields the following explanation E_1.

(20) In (16a) *X* and *Y* are coreferential, because *Y* is bound in ....

So, in the light of cr_1 (i.e. the context-dependent interpretation of the theoretical term *bind*) it is clear that *coreferential* (= T_1) is a relation between two nodes in a constituent structure, in particular, between two noun phrases. Thus, the term T_1 gets the conceptual representation CR_1.

(21) CR_1: Relation between two NPs.

However, (19) is, of course, not the only explication of the term *bind.* Among many other suggestions, Williams (1987) introduced the notion of *theta-binding* which may be conceived of as the lexicalization of the conceptual representation cr_2 of t_1. Consider (22):

(22) *Y* is theta-bound if there is a theta-role *X* which th-commands *Y* and is coindexed with it.

Then, the explanation E_2 is this.

(23) In (16a) *X* and *Y* are coreferential, because *Y* is theta-bound in...

Although E_2 seems to have been motivated by the explanandum Em_1, which is a statement containing the term T_1 with the conceptual representation CR_1 as specified in (16), what is really explained here is no longer Em_1, but a new explanandum Em_2. The reason why this is so is that cr_2 does not select CR_1 from the family of conceptual representations associated with T_1 (the term *coreferential*), but rather, CR_2.

(24) CR_2: Relation between two theta-roles.

Indeed, the relation between cr_1 and cr_2 can easily be identified with that of conceptual shift, as mentioned earlier: whereas in (19) the conceptual representation of t_1 is localized in a theory which captures structural relations between nodes of constituent structure, in (22) it has been shifted to another domain, namely, theta theory. Consequently, cr_2 selected another element of the family of concepts associated with T_1, namely, CR_2, where it is probably conceptual shift again which relates CR_1 and CR_2 via their common semantic representation.

The scope of this example can be easily extended to further cases. To mention one more case, Reinhart (1983) introduces another notion of binding which differs from the previous ones in that binding is seen as a semantic operation of binding variables. Therefore, if t_1 is introduced into the explanans Es_1 with the conceptual representation cr_3 which is the result of identifying the c-command relation with the translatability of a pronoun into a bound variable, then the conceptual representation of the term *bind* will be shifted to the domain of the interpretation of sentences. Accordingly, cr_3 will select CR_3 from the family of concepts associated with T_1.

(25) CR_3: Relation between an operator and a variable.

Thus we obtain exactly what we would expect on the basis of the general structure of theoretical terms and explanations in the light of the metascientific extension of the two-level approach. In particular, the terms in the explanans and those in the explanandum are associated with families of concepts FC_i whose members are related to each other by the conceptual operations conceptual shift or conceptual specification. Consequently, the relationship between the explanans and the explanandum is rooted in the principle of conceptual selection which relates the particular conceptual representations in the former to those in the latter. Therefore, *the explanans may modify the explanandum.*

What remains to be shown is how this finding answers $(P)_{ch5}(b)$. Consider:

- By definition, the explanandum *Em* of a scientific explanation *E* is a description of a fact to be explained whereas the explanans *Es* contains some general statement which should explain this fact. However, as we have just seen, the relationship between the *Em* and the *Es* is based on the principle of conceptual selection, so the latter can modify the former. Consequently, if this modification does indeed take place, then we cannot say that the fact described by *Em* is subsumed under the general statement which the *Es* includes. However, such a subsumption is not excluded at the outset, either: if *Es* does not select a new *Em* differing from the original one — which of course may be the case — then subsumption is possible. Thus we obtain that scientific explanations are *not* necessarily subsumptive.
- If *Es* selects an Em_2 instead of the original Em_1, the relationship between the *Es* and the Em_1 is *not* deductive. However, in two other cases it *may be* deductive. Firstly, when the *Es* does not select another Em_2 instead of Em_1. And secondly, if we consider the relationship between *Es* and Em_2. Accordingly, scientific explanations may be, but are *not* necessarily deductive.
- It is also evident that the structure of scientific explanations we have revealed yields their *predictiveness.* Since *Es* may select Em_2 which may not

> be identical with Em_1 one originally wanted to explain, it is straightforward that Em_2 meets the criteria by which something counts as a prediction. It is related to *Es* in some way, but it is not available at the time at which the latter is available.[10]

Thus the metascientific application of the two-level model yields the following answer to $(P)_{ch5}$ (b): if one accepts the metascientific extension of the two-level approach, then the relationship between the explanans and the explanandum rests on the principle of conceptual selection. It is this aspect which is considered to be *relevant.* Although, by using the traditional terminology, scientific explanations in generative linguistics are — in the light of the two-level approach — not necessarily subsumptive, not necessarily deductive, and *predictive,* it is not these aspects which are focussed on.

5.4 Theoretical terms and pragmatics

The two-level model draws a dividing line between the domain of language and that of communication, where speech acts are construed as interconnecting these two fields in a particular way (see (7) in Section 2.2.1). That is to say, a 'speech act' makes a 'linguistic utterance', on the basis of its meaning, the source of a 'communicative sense'. But the 'communicative sense' of an utterance, according to this approach, is not a linguistic entity at all; rather, it belongs to the domain of 'social interaction'. Let us remind the reader of the notions introduced in (7) in Section 2.2.1 so as to capture this idea (see Bierwisch 1980). There are two crucial points here.

Firstly, in the two-level model there is *no pragmatic component,* because the phenomena, which within the framework of other approaches count as 'pragmatic' in whatever sense, are here distributed between an utterance meaning *m* and a speech act *sa* (Bierwisch 1980) — that is, between the conceptual module and the module of social interaction. This means that, on the one hand some of the phenomena which other approaches treat as part of 'pragmatics' are *clearly captured* by the two-level approach; on the other hand, however, they are captured by subdividing the field of 'pragmatics' into certain modules and, therefore, what matters is *not* the category 'pragmatics' but the specific way the interaction of these modules accounts for the phenomena at issue.[11]

Secondly, according to Bierwisch (1980), *ias* and *cs* belong to the module of social interactions and are, therefore, *not* accessible to the methods of linguistic analysis. That is to say, a communicative sense depends on the interplay between an utterance meaning and the principles or conditions of social interaction determining the interactional setting of a given utterance; but this interplay can be captured only on the basis of a suitable theory of social interaction which, in Bierwisch's opinion, is outside the scope of linguistics:

> [...] we need a theory of the system [...] of rules and structures of social interaction, of which we have vague fragments, at best. But it is a rational task to construct such a theory, albeit not a task that linguists or philosophers of language are competent to accomplish in any serious way. (Bierwisch 1980: 15.)

Thus, with respect to the question of how theoretical terms are related to pragmatic factors, the metascientific extension of this approach yields two conclusions. Firstly, although in the light of the two-level model the structure of theoretical terms is — as a result of the interplay between the semantic and the conceptual representation — characterized by a considerable degree of 'variability' and 'flexibility' (see Section 2.2.1), it does not make sense to ask the question of how 'pragmatic' factors influence this structure. Secondly, the aspects outside this kind of 'variability' and 'flexibility' cannot even be captured by linguistic means.

So, the answer to $(P)_{ch5}$ (c) is that in the framework of the two-level approach the assumption of a pragmatic module seems to be implausible and the nature of theoretical terms is explained by the interaction between relatively autonomous semantic and conceptual factors. This is not to deny the relevance of the phenomena which in a pre-explicative sense may count as 'pragmatic'; however, these phenomena are captured by means which do not make use of the category 'pragmatics'.

5.5 Summary

In sum, the considerations of the last three sections suggest the following answer to $(P)_{ch5}$.

$(SP)_{ch5}$ If the metascientific extension of the generalized modularity hypothesis (MMH′) is accepted, then in generative linguistics

a. the relevant aspect of theoretical terms is that they are semantically underdetermined, because they rest on the relation between their semantic representation and their conceptual representation;

b. in accordance with (a), the relation between the explanans and the explanandum is rooted in the particular mechanisms of conceptual selection which, in turn, is closely related to the particular mechanisms of conceptual shift and/or conceptual specification;
c. in accordance with (a) and (b) the assumption of a pragmatic module is implausible and central properties of theoretical terms should be explained not by using this category, but rather, by the specific interaction between relatively autonomous semantic and conceptual factors.

Even if these findings may seem convincing, the case study carried out touches on certain very serious limits of the way the two-level approach has been applied. Section 14.3 will be devoted to discussing some of them.

Chapter 6

Conclusions

The aim of the chapters in this part of the book has been twofold. Firstly, we examined the extent to which a modular approach and a holistic approach to cognitive semantics yield new tools for the solution of problems traditionally raised by the philosophy of science. As a result, our analyses showed that $(SP)_{ch4}$ and $(SP)_{ch5}$ are well motivated. Secondly, by developing a modular and a holistic approach to the 'cognitive science of science' we set out to give alternative answers to analogous questions within two different frameworks.

A brief comparison of the 'received view' (RV) with $(SP)_{ch4}$ and $(SP)_{ch5}$ suggests that the two-level model and the cognitive theory of metaphor have *common implications:* they reject the hypotheses which the analytic philosophy of science maintains and which were summarized in (RV) in Section 3.6.

Nevertheless, the *means* by which the two models drew this picture are *very different* in several respects:

- The empirical hypotheses — i.e. the generalized modularity hypothesis (MH′) and the generalized holistic hypothesis (HH) — which the two-level model and the cognitive theory of metaphor presuppose are incompatible.
- The way in which their metascientific extensions interpret the structure of theoretical terms and scientific explanations seems to be very different. On the one hand, *the cognitive theory of metaphor*
  - focuses on the metaphorical structure of theoretical terms;
  - emphasizes that metaphors are not to be understood in the narrow sense of individual linguistic expressions of some kind, but rather, in a much broader sense as networks of expressions through which scientific theories conceptualize that segment of the world which they investigate;
  - does not restrict scientific explanations to statements (sentences, propositions), but captures them as the relationship between conceptual domains;
  - infers the main properties of scientific explanations from the unidirectionality of metaphorical processes;
  - treats the semantic, conceptual and pragmatic aspects of theoretical terms and explanations as an indivisible whole.

On the other hand, *the two-level approach*

- attributes to metaphors a marginal role in scientific theories, while it focuses on other kinds of conceptual principles (conceptual selection, conceptual shift and conceptual specification) which operate on the literal use of expressions;
- considers the terms affected by these operations as individual linguistic expressions which enter the explanantia;
- regards explanations basically as sets of statements with a propositional structure connected by certain (conceptual) operations;
- maintains that explanations are governed by the interaction of the explananda and the explanantia;
- implies that the notion of 'pragmatics' cannot be made use of if the structure of theoretical terms and that of explanations have to be captured.

As a result, the similarities and the differences boil down to the following generalization: Since $(SP)_{ch4}$, and $(SP)_{ch5}$ differ considerably from the 'received view' (RV), the conclusion follows that cognitive semantics, applied as a tool for the cognitive science of science, may offer *novel and unexpected solutions to classic problems of the philosophy of science.*

So, the main finding of this chapter thus summarized serves as a plausible argument for the workability of the metascientific extension of cognitive semantics (MECS) introduced in Section 1.3.3. Since, however, the tools by which the two approaches rejected the 'received view' (RV) and suggested $(SP)_{ch4}$ and $(SP)_{ch5}$ are very different, the question arises as to what further implications these *differences* have. Therefore, it is *not* the case that the two approaches can be applied to the *same* problems in the *same* way and to the *same* extent.

Consequently, it is rational to try *to spell out the prospects of both approaches with respect to their different applicability.* In order to do so, we have to find a topic which is both relevant and instructive enough to illustrate these differences.

As recent controversies show, one of the central issues of present day philosophy of science is the question of what kind of relationship exists between what is called the 'conceptual' and the 'social' factors of scientific knowledge. Due to the heated discussions which this problem has given rise to and due to its importance for the philosophy of science, it is ideally suited for testing the workability of approaches to scientific knowledge. Let us, therefore, explore *the possibility of a sociological extension* of the modular and the holistic approach, respectively. Such an extension may be interesting from at least two points of view:

- It may show in what way and to what extent two different versions of cognitive semantics may contribute to the cognitive science of science by capturing one of the *central problems of the philosophy of science*, namely, the relationship between the so-called 'conceptual' and 'social' aspects of scientific knowledge.
- The way these two potential approaches to the cognitive science of science may account for the relationship between the 'conceptual' and the 'social' aspects of scientific knowledge, will lead to *significant consequences* concerning possible accounts of further problems.

In accordance with these expectations, in Part III of the book both the modular and the holistic approach will be tested with respect to the problem of the relationship between the 'social' and the 'conceptual' aspects of scientific knowledge.

Part III

Prospects: Sociological extensions

Chapter 7

The background

For several decades the philosophy of science was considered by working scientists to be an esoteric discipline dealing with highly abstract problems which have little or even nothing to do with the practice of scientific research and the 'real' problems scientists were interested in. However, over the last 30 years, there emerged a problem which gave rise to constant discussions on the nature of scientific knowledge. This problem seems to have changed the philosophy of science into a discipline which may be attractive not only for the proponents of their own field themselves, but also for a wide audience of working scholars and open-minded intellectuals. The problem is this: To what extent and in what way do 'social' factors influence scientific knowledge? The question has a long history, a very rich literature and the ways it has been answered so far constitute a complicated network of possible views of the nature of scientific knowledge. In a very simplified form which cannot stand for a thorough overview, but which will allow us to find a suitable point of departure in this part of the book, the following aspects of the development of the problem should be mentioned:

- The analytic philosophy of science which dominated the scene till the second half of the 20th century centered around the assumption of the 'rationality' of scientific knowledge. Within the analytic philosophy of science there may be different views of what 'rationality' is, but all these views agree on at least two important points. Firstly, scientific knowledge is clearly determined by what is called 'evidence' and some a priori principles which are assumed to govern the way scientific theories account for this evidence. Secondly, although scientists don't always act rationally and are sometimes influenced by 'external' factors such as certain social circumstances, such cases of social influence are to be regarded as anomalies which lead to false, irrational, unacceptable results.
- In the first half of the twentieth century the first contributions to the 'sociology of knowledge' associated with authors like Durkheim, Maus and Mannheim were put forward. Basically, these early works argued for the

claim that in many cases 'knowledge', whatever it may be, is substantially influenced by certain social factors such as interests or authority.

- The more or less coherent view of the 'rationality' of science was questioned by Thomas Kuhn's seminal *The Structure of Scientific Revolutions* as well.[1] Kuhn's notions of 'normal science', 'paradigm', 'scientific revolution', which he illustrated by case studies taken from the history of science, boil down — in a very simplified form and in a specific respect — to the view that scientific research is the result of the interaction between scientific communities, their traditions and their social environments.
- The attempt of the early sociologists of knowledge and Kuhn's work were the source of the rapid and radical development of various new schools in the sociology of scientific knowledge in the seventies and eighties. Although there is not enough space to list them all, the following should be mentioned because of the important role they played in later developments.
- The Edinburgh School of the Strong Programme in the sociology of knowledge, whose outstanding personalities are David Bloor, Barry Barnes and others, (Bloor 1976, 1983, Barnes et al. 1996, etc.) has been the most widely discussed approach in the past 30 years. The Strong Programe is built on four tenets, according to which the sociology of knowledge should be causal, impartial, symmetrical and reflexive.[2]
- Laboratory studies carried out by Bruno Latour and Steve Woolgar, which are rooted, among other things, in the ethnomethodological tradition and are one version of social constructivism, also raised heated controversies. In their seminal case study on the discovery of TRF(H) (i.e. Thyrotropyn Releasing Factor, a Hormone), they show how so-called 'facts' are socially constructed in the laboratory by scientists (Latour and Woolgar 1979).
- The social constructivism of Karin Knorr-Cetina (1981) is important as well. Knorr-Cetina shows that what are regarded as 'facts' and 'truths', are constructed by institutionalized science and that scientific method is not independent of particular circumstances rooted in historically developed social practice.
- Finally, mention must be made of the influential works by the authors of the Bath School like Collins (see e.g. Collins 1985, 1990). Further works by Latour (1987), (1988) gained great importance in the nineties. Brannigan (1981), Pickering (1984), and Shapin and Shaffler (1985) also figure as often-quoted achievements of the sociology of knowledge etc.[3] All these works are basically concerned with the laboratory and, in several respects, spell out the central claim of the sociology of science according to which

scientific knowledge is somehow influenced by society.
- The development sketched in this very simplified manner led to constant debates which sometimes were very heated indeed: "The level of vitriol is greater than in other academic debates, perhaps because a great deal hangs on it politically" (Brown 2001: 449). The sequence of these debates is also called the 'science wars' which culminated in the so-called 'Sokal hoax'. A physicist named Alan Sokal published a 'fictitious' article in a renowned scholarly journal in which he imitated the style of those who question the role of 'rationality' in the sense of the analytic philosophy of science (Sokal 1996a, 1996b, Sokal and Bricmont 1998). This hoax gave rise to very fierce reactions as a result of which the problem of the social factors of scientific knowledge was pushed even more into the spotlight — not only within the relatively small and abstract field of the philosophy of science, but by a much wider audience of sociologists, politicians, working scholars and interested intellectuals (Brown 2000).

The above sketch is intended to indicate that the problem of the social factors of scientific knowledge is one of the *central* issues of contemporary philosophy of science, one which has unquestionable relevance. Therefore, it is ideally suited for testing the potential of any approach which aims at capturing substantial aspects of scientific knowledge.

In this respect a very interesting and fruitful way of sharpening the problem of the social factors of scientific knowledge presents itself, if first we consider two things. On the one hand, we have to remember that what we are after is the contribution of cognitive semantics to the *cognitive science of science* — and *not* to the sociology of knowledge. On the other hand, one of the debates which took place in the course of the process sketched above *sharply contrasted the cognitive science of science with the sociology of knowledge.*[4] Therefore, it would be an interesting and far reaching argument for the fruitfulness of the metascientific extension of cognitive semantics if it could be shown that the two-level approach and the cognitive theory of metaphor are capable of *resolving the alleged conflict* between the cognitive science of science and the sociology of knowledge.

The main feature of the debate mentioned in the previous paragraph is the depth of the antagonism between those who claim the primacy of the 'cognitive' aspects of scientific knowledge and those maintaining that of 'social factors'. For example, Bruno Latour emphasized the need for a ten-year moratorium on cognitive studies of science:

> Any study of mathematics, calculations, theories and forms in general should [...] look at how observers move in space and time, how the mobility, stability and combinability of inscriptions are enhanced, how the networks are extended, how all the informations are [sic] tied together in a cascade of re-representation, and if, by some extraordinary chance, there is something still unaccounted for, *then, and only then, look for special cognitive abilities. What I propose here, as a seventh rule of method, is in effect a moratorium on cognitive explanations of science and technology! I would be tempted to propose a ten-year moratorium.* (Latour 1987: 246–247; emphasis added)

In contrast to this, Slezak as the proponent of artificial intellegence research radically questions the tenability of the sociology of knowledge:

> Although decisive refutations are rare in science, I suggest that this work [artificial intelligence research] has a significance for the strong programme somewhat like the alleged impact of Wohler's synthetic urea on the doctrine of vitalism: *pursuing the original research programme any further is seen to be pointless.* (Slezak 1989: 570; emphasis added)

Consequently, Nickles seems to be right in claiming that "there does not seem to be room in a single universe" (Nickles 1989: 243) for both the sociology of knowledge and the cognitive science of science.

In the next two chapters we will develop case studies which try to show that, although in very different ways, both the two-level approach and the cognitive theory of metaphor can be extended in such a way that they resolve the antagonism between the cognitive science of science and the sociology of knowledge. Nevertheless, as the case studies in part II, also Chapters 8 and 9 serve to exemplify not only the prospects but the limits of the applications, too. These limits will be discussed in Chapter 14.

In the case studies, due to the basic differences between the two-level approach and the cognitive theory of metaphor, we will choose two different points of departure. In Chapter 8 it will be Bloor's Strong Programme whose integration with the two-level approach will be attempted. In Chapter 9 one specific version of ethnomethodology which led, among other things, to the development of a sociolinguistic trend, namely, conversation analysis, will be integrated with the cognitive theory of metaphor. In Chapter 10 some of the far-reaching implications of the case studies carried out in Parts II and III will be discussed.

Chapter 8

Case study: A sociological extension of the modular approach

8.1 Introduction

8.1.1 The problems

That the object of the case study to be carried out is *generative linguistics*, is not the result of an arbitrary decision. In the debate between cognitivists and sociologists mentioned in Chapter 7 the example of generative linguistics played an important role. Slezak takes Chomsky's work as an example and asks a series of rhetorical questions:[1]

> To what social causes can the detailed content of Chomsky's transformational grammar be attributed? Has the radical shift in Chomsky's theoretical views been as a result of changed social circumstances? (Slezak 1989: 587)

Therefore, we restrict the basic question of this chapter to generative linguistics which serves as the paradigmatic example of how certain 'conceptual' and 'social' factors of scientific knowledge are related.[2] Answering Slezak's challenge concerning generative linguistics seems to provide a good opportunity to test the workability of an approach which claims to take account of both certain 'cognitive' and 'social' aspects of scientific knowledge in a proper way. What we need then, is an appropriate extension of the modular approach we outlined in the previous chapters. Thereby, in analogy to the case studies in Chapters 4 and 5 we must not argue for the acceptance of the findings which the application of the two-level approach yields; rather, our concern with the prospects and limits of cognitive semantics suggests that we simply have to explore what the case would be if the two-level approach were accepted. This would amount to raising the following problem: "What kind of relationship would there be between certain 'conceptual' and 'social' factors of generative linguistic theories, *if* the two-level approach as an example of modular cognitive semantics could be extended so that it captures the relationship between certain 'conceptual' and 'social' aspects of theories?" Nevertheless, since in the following argumentation much depends on whether the two-level approach can be extended sociologically, and if so, how, it is important to split up this problem into two questions and answer them separately. The first is:

$(P1)_{ch8}$ What kind of relationship would there be between certain 'conceptual' and 'social' factors of generative linguistic theories, if the metascientific extension of the generalized modularity hypothesis (MMH′) were accepted?

$(P1)_{ch8}$ can be answered after we have raised and solved $(P2)_{ch8}$:

$(P2)_{ch8}$ Can the two-level approach as an example of modular cognitive semantics be extended so that it captures the relationship between certain 'conceptual' and 'social' factors of scientific knowledge?

These two problems will be tackled by integrating some of the basic aspects of Bloor's Strong Programme with the modular approach. Befor doing so, background knowledge of the Strong Programme has to be provided.

8.1.2 The Strong Programme for the sociology of knowledge

In the late seventies and early eighties the analytic philosophy of science was challenged not only by Quine's programme but also by the sociology of knowledge. Within the latter, the Strong Programme for the sociology of knowledge as elaborated by David Bloor and his Edinburgh School put forward especially radical theses.

The Strong Programme and Quine's naturalism share the view that, in opposition to the theses of the analytic philosophy of science, scientific knowledge is not governed by the universal and a priori principles of rationality and that the philosophy of science should be treated as an 'empirical' discipline.[3] However, the Strong Programme differs from Quine's intentions among others in that it does not trace back scientific knowledge to psychological factors, but rather, to social ones. In spite of this difference the Strong Programme is to be considered as one current manifestation of naturalism — as is the cognitive science of science, in spite of its differences from some relevant aspects of Quine's programme. For example, in his review of the second edition of Bloor's seminal *Knowledge and Social Imagery* Steve Fuller wrote this.

> [...] I plan to explore the implications of regarding Bloor (ibid) as a research program in '*naturalized epistemology*' (Fuller 1993a: 159).[4]

> In conclusion, not only is Bloor a naturalized epistemologist, but he is an *especially consistent naturalist* [...]. (Fuller 1993a: 168)

We may summarize the main characteristics of the Strong Programme in a way which shows how it may be subsumed under the general properties of the naturalized philosophy of science and how it differs, accordingly, from the analytical philosophy of science (see (13) and (14) in Chapter 1).

(1) The Strong Programme for the sociology of knowledge
  a. is *not* a philosophical discipline; rather, it should apply the *scientific* methods of *sociology*;
  b. proceeds, accordingly, in *an empirical, a posteriori* manner;
  c. aims at the *description and explanation* of scientific inquiry by using the a posteriori/empirical methods of sociology in accordance with (a) and (b).

This simple characterization can be be further specified as follows.

Firstly, the Strong Progamme assumes the pluralism of scientific knowledge which it considers to be rooted in social variation:

> Men's ideas about the workings of the world have varied greatly. This has been true within science just as much as in other areas of culture. *Such variation forms the starting point for the sociology of knowledge and constitutes its main problem.* What are the *causes* of this variation, and how and why does it change? The sociology of knowledge *focuses on the distribution of belief and the various factors which influence it.* For example: how is knowledge transmitted; how stable is it; what processes go into its creation and maintenance; how is it organised and categorised into different disciplines or spheres? (Bloor 1976: 3; emphasis added)

Secondly, the sociology of knowledge should therefore be regarded not as a philosophical, but rather, as an a posteriori, empirical discipline which intends to provide causal explanations for the nature of knowledge on empirical grounds:

> The overall strategy has been to link the social sciences as closely as possible with the methods of *other empirical sciences.* In a very orthodox way I have said: only proceed as the other sciences proceed and all will be well. [...] Indeed the central themes of this book [...] have all the characteristics of *straightforward scientific hypotheses.* (Bloor 1976: 141; emphasis added)

Thirdly, Bloor's approach rests on his famous four tenets:

> The approaches that have just been sketched suggest that the sociology of scientific knowledge should adhere to the following four tenets. In this way it will embody the same values which are taken for granted in other scientific disciplines. These are:

> 1. It would be causal, that is, concerned with the conditions which bring about belief or states of knowledge. Naturally there will be other types of causes apart from social ones which will cooperate in bringing about belief.
> 2. It would be impartial with respect to truth and falsity, rationality or irrationality, success or failure. Both sides of these dichotomies will require explanation.
> 3. It would be symmetrical in its style of explanation. The same types of cause would explain, say, true and false beliefs.
> 4. It would be reflexive. In principle its patterns of explanation would have to be applicable to sociology itself. Like the requirement of symmetry this is a response to the need to seek for general explanations. It is an obvious requirement of principle because otherwise sociology would be a standing refutation of its own theories.
>
> These four tenets, of causality, impartiality, symmetry and reflexivity define what will be called the strong programme in the sociology of knowledge. They are by no means new, but represent an amalgam of the more optimistic and scientistic strains to be found in Durkheim (1938), Mannheim (1936) and Znaniecki (1965). (Bloor 1976: 4–5)

The following comments on the four tenets should illuminate their significance.

According to the *thesis of causality*, the philosophy of science should be scientific in that it should not have as its main concern the justification of scientific knowledge on the basis of the a priori principles of rationality, but rather, it should explain and describe knowledge by referring to the causes which might have brought it about. Although the thesis of causality is not compatible with the aprioristic and normative attitude of the analytical philosophy of science in the strict sense, once one has accepted that the science of science may proceed like science does, its plausibility is beyond doubt. If scientific reason is based on revealing the causes of what we find in the world, then there is no denying that it is well-motivated to reveal the causes of what we find in science as well. The same applies to *the tenet of reflexivity*, too: of course, it is acceptable that we must make use of reason when thinking about reason. In addition, the reflexivity thesis attributes to sociology a designated position in that it requires that sociology be both the object of the investigation and the means of metascientific reflexion (see also Brown 2001).

It was particularly the thesis of impartiality and the thesis of symmetry that raised very heated discussions in the literature. The thesis of impartiality and the thesis of symmetry are quite radical: for some commentators radically fruitful and progressive, for others radically unacceptable. On the one hand, some philosophers of science evaluated them as an important turn and assumed that they may lead to the emergence of a new paradigm in the philosophy of science (Hesse 1980, 1988, Fehér 1988). On the other hand, those who could not give up the traditional conception of the general validity of rationality as an essential value of scientific knowledge rejected the Strong Programme vehemently. Very different scholars belonged to this second camp: not only the proponents of the analytical philosophy of science, but also some leading sociologists and proponents of different trends within the naturalized philosophy of science.

The *thesis of impartiality* says that the sociology of knowledge should be interested not only in the positive achievements of science — truth, success, rationality —, but also in what failed or what has turned out to be irrational or false; this is, of course, in sharp contrast with traditional assumptions concerning the nature of metascientific reflection. Most philosophers of science could, for instance, accept the fact that social factors can play some role in shaping scientific knowledge, but their scope must be restricted to false, unsuccessful and irrational assumptions. In contrast, Bloor's *thesis of symmetry,* which requires that true and rational belief should be explained by the same type of causes that underlie irrational, false and unsucessful belief, is wholly unacceptable not only for the proponents of the analytical philosophy of science — it is equally unacceptable even for those who maintain some kind of naturalized philosophy of science, but who do not want to give up the ideal of rationality as a generally valid property of scientific knowledge. Nevertherless, Bloor emphasizes that it is not the intention of the Strong Programme to deny rationality in general by insisting on the much-debated symmetry thesis. Rather, the Strong Programme assumes a different kind of rationality according to which 'rationality' is a social product: in different historical periods there have been different 'rationalities' and, moreover, in different communities different 'rationalities' may be accepted. This is, of course, a relativistic stance, which Bloor deliberately and consciously maintains.

The fierce reactions to Bloor's programme are best exemplified by Larry Laudan's verdict which condemned the Strong Programme "the pseudo-science of science" (Laudan 1984). Those who think that the symmetry thesis and the impartiality thesis allow a progressive and novel approach to the nature of scientific knowledge interpreted this reaction and similar ones as symptoms of a 'crisis' in the sense of Kuhn, and, consequently, attributed a revolutionary role to the Strong Programme in the philosophy of science (see e.g. Fehér 1988, Hesse 1988).

As mentioned in Chapter 7 and Section 8.1., in the present chapter we will contribute to the debate which took place between the sociologists of knowledge and the proponents of artificial intelligence research. The initial question of the debate asked if scientific knowledge was shaped by cognitive *or* social factors. However heated the debate was, its outcome did not seem to be fully destructive, because several contributors maintained that models of scientific knowledge should be elaborated which are capable of accounting *both* for the cognitive and the social factors underlying that knowledge. One of the models which may be fruitful in this respect is an integrated philosophy of science based on the metascientific extension of the generalized modularity hypothesis as elaborated in Kertész (1993). At this point it is important to emphasize that the reconstruction of some main ideas of Bloor's Strong Programme for the sociology of knowledge within the framework of a modular approach to the cognitive science of science seems to be well-motivated and must not be regarded as an arbitrary integration of very different or even incompatible theories. In fact, the generalized modularity hypothesis (MH′) introduced in Section 2.2.1, according to which human cognitive behaviour is organized modularly, is compatible with the Strong Programme. By mentioning the example of 'perception', the Strong Programme declares that *it considers sociological analysis to be compatible with the assumption of modularity*:

> It is perfectly possible and empirically plausible to accept the claim that perception is to a large extent 'modular' — that is, isolated from other components of our cognition, and only influenced by them in a limited way. This in no way rules out a sociological analysis. (Barnes et al. 1996: ix)

At this point two important remarks are in order. Firstly: as the last quotation witnesses, the Strong Programme itself does not exclude the possibility of *integrating the modular approach to scientific knowledge with certain aspects of the Strong Programme*, however controversial the Strong Programme is and however radical some of its claims are. Secondly: the two-level approach doesn't exclude such an integration, either. As we saw in Section 2.2.1 in general and in (17) in Section 2.2.1 in particular, the two-level approach may be conceived of as a complex approach to human cognitive behaviour which *strives to integrate several different theories, and among these, also sociological ones*. This means that also from the point of view of the two-level approach certain findings of the Strong Programme may be reconstructed within the modular framework outlined in Chapter 2 and related to those conceptual factors which the latter is intended to capture at the outset.

This point of departure suggests the following line of argumentation. In Section 8.2 the premisses of the argumentation in the case study will be outlined. It will be shown that the modularity hypothesis can integrate certain 'social' and 'conceptual' aspects of scientific knowledge in a very straightforward way and that this yields the research strategy of a 'modular approach to the cognitive science of science'. From this research strategy we will obtain an affirmative answer to $(P2)_{ch8}$. Section 8.3 is devoted to the application of this strategy to capturing the differences between two types of explanation in generative linguistics. As an answer to $(P1)_{ch8}$ it will be shown that the peculiarities of grammatical explanations rest on the relationship between what are called 'social' and 'conceptual' factors of scientific knowledge and that this relationship is to be explicated in terms of the 'parametrization relation' between universal principles of behaviour belonging to two different modules. Finally, Section 8.4 is a summary of the conclusions.[5]

8.2 The solution to $(P2)_{ch8}$

8.2.1 Empirical hypotheses

To find a suitable point of departure, we have to remind the reader of those aspects of the two-level approach which we will consider as *the starting premises* of the argumentation to follow.

First of all, in Section 2.2.1 we discussed the generalized modularity hypothesis (MH′) which says that human cognitive behaviour is based on the interaction of relatively autonomous subsystems called modules. From (MH′) we obtained the metascientific extension of the generalized modularity hypothesis (MMH′) according to which scientific knowledge in particular is based on the interaction of modules as well. Among other things, the following theses can be inferred from some of the main tenets of the two-level approach (see (1), (2) and (3) in Section 2.2.1) and the metascientific extension of the generalized modularity hypothesis (MMH′):

(MMH′1) (ia–e) in (1) in Section 2.2.1 are candidates for modules underlying scientific behaviour.

(MMH′2) Along the lines of (2) in Section 2.2.1 and (MMH′)

a. these modules consist of universal principles which determine rules and these rules, in turn, determine representations;
b. instances of scientific behaviour are conceived of as sets of representations.

(MMH′3) In accordance with (MMH′) and (3) in Section 2.2.1 a universal principle UP_1 belonging to a module M_1 underlying scientific behaviour is parametrized with respect to (or is a parametrization of) a universal principle UP_2 belonging to a module M_2 if and only if the value of the free parameter P_1 associated with UP_1 depends on the value of the parameter P_2 associated with UP_2.

In the light of $(P1)_{ch8}$ it is (MMH′1) that deserves special attention. In Section 2.2.1 we mentioned a hypothetical list of modules which for example Bierwisch (1980), Grewendorf et al. (1987), Lang and Carstensen (1990), and Lang et al. (1991) assume.[6] Now, the following question arises that is very important both for $(P1)_{ch8}$ and $(P2)_{ch8}$: Which of the possible modules referred to in (MMH′1) correspond to what have generally been called the 'cognitive' and the 'social' factors of scientific knowledge, respectively? Given the list of possible modules, the only candidate which broadly covers the phenomena called 'conceptual', 'cognitive', 'psychological', 'mental' etc. factors from different perspectives of the debate is the *conceptual module*; therefore, we explicate the term 'system of conceptual / cognitive / psychological / mental etc. factors' as 'the conceptual module.' (See also Section 14.3 on some of the problems this suggestion touches on.)

The obvious candidate for explicating 'the system of social factors' is the social module characterized in (i)(d) in (1) in Section 2.2.1. Nevertheless, the social module is assumed to consists of two submodules: namely, the *submodule of social interactions*, and the *motivational submodule* which organizes the objectives, interests and intentions of individuals and groups of individuals. Which of these two submodules should provide a possible explication of the pre-explicative notion of 'social factors of scientific knowledge'? We may find the answer to this question if we have a closer look at the pre-explicative use of the term 'social'. Although in the literature on the sociology of knowledge it is far from clear what is actually meant by 'social', Bloor tends to focus, among others, on *social interests*; see e.g. the very influential works Bloor (1976), (1983).

As witnessed by (i)(d) in (1) in Section 2.2.1, within the frames of the modular approach it is *the motivational submodule* which comprises mental representations of social interests, needs etc.:

> [...] the motivational system which considers our instances of behaviour to be organized by structured hierarchies of aims and partial aims whose highest

> dominating elements consist in fundamental, phylogenetically developed needs and inclinations (Grewendorf et al. 1987: 37; my translation, A. K.)

Consequently, it seems to be plausible to choose the motivational module as a possible explication of the notion 'social factors of scientific knowledge' within a modular approach to the cognitive science of science. (See Section 14.3 on certain difficulties with respect to this decision.)

Given this interpretation of 'conceptual' and 'social' as 'conceptual module' and 'motivational module', respectively, the next question is how the *relationship* between the 'conceptual' and the 'social' factors of scientific knowledge can be explicated. The key to the answer is the notion of *parametrization* which is at the heart of possible interactions between modules in general and which was characterized in (MMH′3). Accordingly, the assumption that social factors contribute significantly to scientific knowledge means that they do so by the value of a parameter. Thus, from (MMH′3) we obtain the thesis which makes the present modular approach appear to be a kind of *integrated approach to the cognitive science of science* in that it is capable of capturing the interaction of the motivational (i.e. one subdomain of the social) and the conceptual aspects of scientific knowledge:

(IPS) The universal principles of the conceptual module underlying scientific knowledge are parametrized by the universal principles of the motivational module.

(IPS) is an explication of the idea that the 'content' of scientific knowledge is influenced by 'social' factors because it claims that this influence is nothing but the value of a parameter which a certain conceptual principle acquires from a motivational one. In sum, what we have said boils down to this:

- (MMH′) and (IPS) are the two empirical hypotheses of a modular approach to the cognitive science of science which can capture the interaction of the 'social' and 'conceptual' factors of scientific knowledge.
- Accordingly, what we have to examine is the interaction of modules underlying generative linguistic inquiry along the lines of the empirical hypothesis (IPS).

Before showing how such an analysis can be carried out, let us overview the main steps of the research strategy of a modular approach to the cognitive science of science which the theses and notions introduced so far suggest.

8.2.2 The research strategy of a modular approach to the cognitive science of science

(MMH′) and its consequences discussed in the previous section straightforwardly yield the sub-problems into which $(P1)_{ch8}$ (i.e. the problem of the relationship between the motivational and the conceptual factors of knowledge) has to be subdivided within the frames of a modular approach to the cognitive science of science. These sub-problems are as follows:[7]

(2) a. Let us assume that a certain philosopher of science (or a group of philosophers of science) aims at a metascientific description and explanation of instances of objectscientific behaviour in a given domain *DO*. Let us identify such an instance with a particular objectscientific theory TO_i. We also know from (MMH′2) that instances of objectscientific behaviour consist of representations *R*, the elements of which are determined by the rules and principles of different modules. Since (MMH′1) suggests a tentative list of these modules, we may assume that the behaviour instance TO_i consists, for example, of a conceptual, a grammatical, a motivational etc. representation. It is these representations which can be described and explained metascientifically, and one can only account for the behaviour instance TO_i by describing and explaining these representations.[8] Accordingly, the explananda of metascientific explanations of instances of objectscientific inquiry are descriptions of representations of instances of objectscientific behaviour. In the light of our central problem we must assume that what has to be explained metascientifically is the conceptual representation C_i of TO_i. Thus, the task to be carried out during the first stage is the identification of the explanandum of such a metascientific explanation. In the present case this boils down to formulating the question: *Why does the instance of behaviour* TO_i *have the conceptual representation* C_i*?* It is this question whose answer should finally be given by the application of a modular approach to the cognitive science of science.

b. We know from (MMH′) that instances of objectscientific behaviour are determined by the interaction of modules. Therefore, the question arises: The interplay of which modules underlies C_i? On the basis of (IPS) we have to assume that it is at least the conceptual and the motivational module that determine C_i.

c. It is also clear that both of these modules consist of universal principles which are motivated independently of each other. Therefore, the next question to be answered is which universal conceptual and motivational principles can be revealed. Let us assume that our philosopher of science has found a conceptual principle UP_C and a motivational principle UP_M.

d. In accordance with (MMH′3) we know that the interaction of modules is primarily, although not exclusively, manifested in the parametrization relation between the principles of the corresponding modules. Therefore, the next step must be to decide which parameters can be associated with the principles that have been detected in (2c), i.e UP_C and UP_M.[9] As a possible solution we may find a conceptual parameter P_C and a motivational one P_M.
e. Now, we have to discover which values the free parameters are set at. In accordance with (IPS) the main assumption is that the conceptual principle UP_C has been parametrized by the motivational principle UP_M which means, in other words, that the value of the parameter P_C associated with UP_C has been fixed in such a way that it depends on the value of the parameter P_M associated with UP_M. (Revealing the value of the parameters and thus revealing the parametrization relation between UP_C and UP_M will yield the explanans of the metascientific explanation we are looking for; see (g))
f. We know that, in accordance with (MMH′2) and (MMH′3), if the free parameters of a given universal principle have been fixed, then a rule is obtained. Accordingly, the next question to be answered after having carried out step (2)(e) is which conceptual rule can be obtained by fixing the parameter P_C. Let us, for the sake of simplicity, assume that P_M was set at the value *a* and this value was transferred somehow to P_C. Then we get a conceptual rule R_C whose constituents are identical with those of UP_C, except that, instead of the open parameter P_C, the value *a* occurs. This rule is one of those which are assumed to govern behaviour in the domain *DO*.
g. Finally, R_C yields the representation C_i immediately. Consequently, the main task has been solved and a metascientific explanation of the conceptual aspect of the instance of objectscientific behaviour TO_i has been proposed. This explanation is *the answer to the initial why-question asked in (2a).*

8.2.3 Conclusions

On the basis of what has been said in Sections 8.2.1 and 8.2.2 we straightforwardly obtain the following solution to $(P2)_{ch8}$.

$(SP2)_{ch8}$ The two-level approach as an example of modular cognitive semantics is capable of being extended so that it captures the relationship between certain 'conceptual' and 'social' factors of scientific knowledge, because

a. it presupposes the two empirical hypotheses (MMH′) and (IPS), and
b. these, in turn, immediately yield the research strategy in (2).

Now we may turn to applying the research strategy to solve $(P1)_{ch8}$, that is to achieve a metascientific explanation of the way conceptual and motivational factors interact in shaping the content of objectscientific explanations in generative grammar.

8.3 The solution to $(P1)_{ch8}$

8.3.1 Background assumptions

As is well known, there are considerable differences between Chomsky's Standard Theory of Generative Grammar (1965) and the Government-Binding Theory (Chomsky 1981, 1986, 1987). The differences are, among other things, evidenced by the structure of the objectscientific explanations whose aim is to account for the structure of natural language sentences. They affect both the explananda and the explanantia of grammatical explanations. The core of these differences is, firstly, that whereas the Government-Binding Theory accepts one particular version of the modularity hypothesis (MH) which says that knowledge of language is organized in a modular fashion, nothing like this was maintained in the Standard Theory.

Secondly, the difference between rule-oriented and principle-oriented explanations is vital as well. The Standard Theory sets out to find grammatical rules which explain particular phenomena. For instance, if a linguist wants to explain the structure of a passive sentence in English, then he/she has to find a rule which is specific to this phenomenon, i.e. a passive transformation. In contrast to this, Government-Binding Theory is expected to deliver a very different kind of explanans. In particular, there are assumed to be no construction-specific grammatical rules which are responsible for the structure of certain constructions like the passive, but rather, the explanation is obtained by uncovering universal principles of language together with the free parameters associated with them. Once the parameters are fixed, one obtains either the rules of a particular language or the representations themselves which one wanted to explain. Nevertheless, neither the universals nor the language-specific rules are construction-specific; in Government-Binding Theory there is nothing like a rule of passive constructions. As we see, the most interesting account of the differences between the two types of explanation is that whereas those of the Standard Theory are *rule-oriented*, the explanations of the Theory of Government and Binding are *principle-oriented.*[10]

Thirdly, it is important to realize that the difference between rule-oriented and principle-oriented explanations indicates, among other things, a difference between the *content* of the two theories. For, what the Standard Theory is about, is, in accordance with its rule-oriented strategy of explaining linguistic phenomena, a set of assumptions concerning construction-specific rules of, say, English: rules like 'passive-transformation' and the like. Government-Binding Theory says that it is not sufficiently general to reach descriptive/explanatory adequacy to accept that there exists a rule to be called 'passive transformation'. Rather, the peculiarities of passive sentences are described as resulting from the interplay of certain universal principles of language (see, for example, Chomsky 1991: 24).

Thus, the Standard Theory and Government-Binding Theory may deliver *different kinds of empirical hypotheses* concerning the regularities which are supposed to underlie a native speaker's knowledge of language. Although in the controversy between cognitivists and sociologists in the philosophy of science it is far from clear what is meant by 'content',[11] intuitively it is plausible to maintain that if two theories consist of different empirical hypotheses about the segment of the world which they have to investigate, then they differ in 'content'.

Summing up what has been said so far, the following claims have been made.

- The explanations of the Standard Theory are rule-oriented and those of Government-Binding Theory are principle-oriented.
- As a result of this, the 'content' of the two theories may also be different. Given this, the question for us is how a modular approach to the cognitive science of science can propose *a metascientific explanation* of the fact that the content of the Standard Theory may be different from the content of Government-Binding Theory.

Therefore, we have to follow the steps of the research strategy outlined in (2).

8.3.2 Step (2a)

In the sense of step (2a) let us assume that a philosopher of science PS has observed two instances of objectscientific behaviour *x* and *y*. *x* is an explanation put forward by a representative of the Standard Theory and is, accordingly, rule-oriented. *y* is a principle-oriented explanation proposed within the framework of Government-Binding Theory. Due to (MMH'2) PS presumes, then, that both instances of behaviour correspond to a set of representations such as the conceptual, the motivational, and the grammatical. He/she describes these representations with the help of a certain formalism, whose details do not matter now, and chooses the description of the conceptual representation C_x of *x* and the description of the conceptual representation C_y of *y* as the explananda of a metascientific explanation. The question, then, is, *why x has the conceptual representation* C_x *and why y has the conceptual representation* C_y. In other words he/she has to provide a metascientific explanation of C_x and C_y.

8.3.3 Steps (2b) and (2c)

According to steps (2b) and (c) this question can be answered by revealing the principles and parameters of at least the conceptual and the motivational module underlying the objectscientific explanation at issue. Therefore, the next problem to be solved is what universal conceptual and motivational principles may be responsible for C_x and C_y.

As regards possible conceptual principles governing scientific behaviour, we have argued at length elsewhere for the assumption that in the case of scientific theory formation in general, one version of the *relevance principle* is valid (Kertész 1991). The intuitive idea of the relevance principle goes back to Sperber and Wilson (1986) but should be slightly modified if applied to scientific explanations. In particular, I suggested the following reconstruction of the idea of the relevance principle as a possible universal of scientific behaviour:

(T_C) Every explanation is put forward in a context *ct* under the assumption of its maximal relevance with respect to *ct*.

Providing a proof of the tenability of this principle would be beside the point now.[12] If we hypothetically assume the plausibility of (T_C), then what matters in the present case is simply how it may lead to a possible explanation of C_x and C_y. Since we have presupposed that conceptual principles interact with motivational ones, in order to answer this question the philosopher of science PS must set out to find a principle of the motivational module, too. He/she may obtain a possible principle by trying to reconstruct Bloor's considerations concerning *models*. The following quotation should serve as a point of departure:

> It is implicit in the very idea that the patterns of objects which are within the reach of our experience can function as models. For consider how models work and what happens when one piece of behaviour is modelled on another. The result is precisely to detach the derivative behaviour from that on which it is modelled. Think here of the carpet weavers. A weaver picks up the way that the pattern goes by watching and working with others. He can then function autonomously and apply and reapply the technique to new cases. He could, for example, set out to weave a carpet bigger than he had ever seen anyone weave before, but need only have learned and practiced on small ones. (Bloor 1976, p. 90)

One claim which this quotation makes is that 'models' are always connected to material objects embedded in a certain situation. Another is that, once a given set of properties of such objects are selected as 'models', the latter can be extended to new situations which are not immediately experienced. This role of models raises three questions. Firstly, what are 'models'? Secondly, if such models are, according to Bloor, tied to specific situations, material objects or persons, how can they be transferred to new situations? And thirdly, what does the choice of a given 'model' depend on?

As regards the first question, Bloor does not define his notion of 'model'. But if we try to reconstruct the main tenets of his approach within the motivational module of a modular approach to the cognitive science of science, then we should interpret 'models' roughly in the sense of 'mental models' as characterized, for example, in several writings of Johnson-Laird and Bierwisch (see Johnson-Laird 1980, 1983, Bierwisch 1981, 1983a); however, the details do not matter here. In any case, if in a modular approach to the cognitive science of science it is legitimate to explicate Bloor's notion of 'model' as 'mental model', then it has at least the following two aspects:

- Mental models are results of the fact that humans come into contact with their material environment and represent mentally the experiences gained thereby.
- Mental models are the conceptual representations of situations which one experiences and rest both on sensory and inferential information. In particular, the mental models of material objects are their *conceptual representations.*

However superficial these remarks are, we may now consider the second and third question mentioned above. Bloor's answer to the second question is that it is analogies which transfer 'models' originally connected with material objects to situations which are not immediately experienced. But, if we explicate 'models' as 'mental models', then this means that there must be *conceptual principles* which govern such transferences. Therefore, it is these conceptual operations which facilitate the extension of 'models' to new situations.

As regards the third question, Bloor's 'models' are closely connected to material objects embedded in situations in which the members of a scientific community operate. A material object obviously has many properties which are conceptually represented in the mind. However, not all of these conceptual representations can function as 'models' (i.e. 'mental models') in our sense. Now, what Bloor also maintains is that the choice of these models is governed by *social interests.* In a modular framework, social interests, whatever they may be, belong to the motivational module. This means, in our terminology, that, on the one hand, the extension of 'models' to new situations is a conceptual process governed by *conceptual principles.* On the other hand, the choice of the initial model (which is extended in such a way) rests on principles belonging to the *motivational* module. Which properties of a given object should be chosen as a 'model' and be extended to new situations depends on what counts as important, useful etc. in a given society. In what follows those conceptual representations of objects which are the starting points for conceptual extensions will be called *I-models* ('initial models'). I-models are, however, conceptual representations of certain properties of material objects. Therefore, those properties of objects whose conceptual representation functions as an I-model, will be called *I-properties.* So, Bloor's claim that the choice of models is governed by social interests, leads in our framework to the following assumption:

(T_M) Social interests select I-properties.

Since in Bloor's theory the corresponding thesis is accorded general validity, we should assume that (T_M) is a *universal principle of the motivational module.*

These considerations show that the tenets of Bloor's 'Strong Programme' for the sociology of knowledge can be reconstructed in a modular approach to the cognitive science of science. The result of such a reconstruction yields the theses of the motivational sub-theory of a modular approach to the cognitive science of science which is a metascientific theory of the social aspects of scientific knowledge. This motivational sub-theory should then be integrated with the conceptual sub-theory which contains, among other things, descriptions of conceptual principles such as (T_C). But such an integration will only be possible if it is not only the universal principles of the conceptual and the motivational module which are described by the corresponding subtheories, but also the free parameters associated with the principles. Therefore, the next question is what these parameters are (see step (2d)).

8.3.4 Step (2d)

Following (MMH′3) we know that universal principles may be associated with free parameters. We also know that the interaction of two modules is manifested, among other things, in the fact that the value of a parameter associated with a certain principle belonging to one module is set by the value of one parameter associated with another principle belonging to another module. Consequently, the next step consists in revealing such a parameter.

As regards the universal principles of the conceptual module, it is the *context* which plays the role of an open parameter; see (11) in Section 2.2.1 and Bierwisch (1983b: 79). The reason for this is simple: the context (conceived of as a conceptual entity) is always a *variable* whose value may change from case to case. Thus in the case of (T_C) the open parameter is the context *ct.* (T_M) contains a free variable as well: namely, *social interests.* Although in the literature on the sociology of knowledge it is far from clear what count as social interests, it goes without saying that interests are not constant in so far as they are exposed to changes and vary considerably. Having identified the parameters associated with the two principles, we must next find out — along the lines of step (2e) — how these parameters are fixed.

8.3.5 Step (2e)

According to (IPS) the central empirical hypothesis of a modular approach to the cognitive science of science is that the crucial principle of the conceptual module, in our case (T_C), is a parametrization of a principle of the motivational module which is, in the present case, (T_M). Therefore, we must assume that the value of the conceptual parameter *ct* 'depends' on the value of the motivational parameter 'social interests'. Accordingly, the question is, which social interests governed research in the Standard Theory and in Government-Binding Theory, respectively.

Metascientific investigations carried out, among others, by Riley (1987) and Forrai (1987) clearly show that the two theories follow different interests which are, however, not difficult to identify.[13] Forrai and Riley argue independently that the primary interest which guides research in the Standard Theory is 'systematic description of grammar'. Let us label this particular value of the interest parameter of the principle (T_M) p_M. In contrast to this, in Government-Binding Theory the value of the interest parameter is 'strong restriction of grammar'; let this be denoted by q_M.

At first sight, the interests at issue appear to be 'cognitive interests' and this, if true, would trivialize our analysis. However, this is not the case, because they are essentially social in nature. For example, 'systematic description of grammar' reflects the fight of the early generativists to develop a linguistic theory which meets the highest standards of the methodolgy of natural science. The fulfillment of these standards is, of course, closely connected with authority, social position and power (see also Newmeyer 1986, 1991). Likewise, 'strong restriction of grammar' is not independent of social factors either, as is witnessed by the world-wide success of Government-Binding Theory not independent of the fact that its proponents gained authority, social positions, and power. (One possible reason for preferring strong restriction of grammar is that in this way all the phenomena could be explained which were the subject matter of the Standard Theory and, in addition a series of further phenomena (e.g. language acquisition, similarities and differences between particular languages etc.) which the Standard Theory could not account for.)

8.3.6 Step (2f)

Consequently, we have a motivational principle, namely (T_M) and two values, i.e. p_M and q_M, at which the parameter in (T_M) can be fixed. Due to (2e), the possible values of the conceptual parameter *ct* must depend on p_M and q_M. Therefore, *ct* may have two values, namely, p_C and q_C.[14]

Now, it is well-known that in the Standard Theory Chomsky interprets 'systematicity' as 'simplicity'. Consequently, in the case of the Standard Theory, giving the context-parameter *ct* the value p_C leads to the following rule of research:

> (R_{ST}) Explanations are determined under the assumption of their maximal simplicity.

Conversely, in Government-Binding Theory setting the context-parameter associated with (T_C) at the value q_C yields the following rule:

(R_{GB}) Explanations are determined under the assumption of their maximal restrictiveness.

Accordingly, step (2f) has been carried out.

8.3.7 Step (2g)

What remains now is to derive the conceptual representation C_x of the instance of scientific behaviour *x* and the conceptual representation C_y of *y*. As is commonplace, (R_{ST}) is the primary criterion for evaluating explanations in the Standard Theory resulting in rule-oriented explanations and (R_{GB}), in turn, serves to evaluate explanations in Government-Binding Theory leading to principle-oriented explanations. Thus, (R_{ST}) and (R_{GB}) immediately yield C_x and C_y, respectively. Consequently, we have put forward a metascientific explanation for the difference between C_x and C_y. The two metascientific explanations say:

(E_{ST}) *x* is a rule-oriented explanation in generative linguistics, because its conceptual representation C_x is determined by the rule (R_{ST}). (R_{ST}), in turn, results from the parametrization-relation between the universal principle (T_C) of the conceptual and the universal principle (T_M) of the motivational module. This parametrization relation consists in the fact that the free parameter *ct* associated with (T_C) was set at the value p_C whose origin is (T_M).

(E_{GB}) *y* is a principle-oriented explanation in generative linguistics, because its conceptual representation C_y is determined by the rule (R_{GB}). (R_{GB}), in turn, results from the parametrization-relation between the universal principle (T_C) of the conceptual and the universal principle (T_M) of the motivational module. This parametrization relation consists in the fact that the free parameter *ct* associated with (T_C) was set at the value q_C whose origin is (T_M).

Thus (E_{ST}) and (E_{GB}) yield the *answer* to our initial why-question asked in Section 8.3.2.

8.3.8 Conclusions

Although, due to the programmatic nature of the present considerations, this case study dispenses with many minor details which could have made it more precise, it still shows that the result is exactly what could be expected at the outset:

- We revealed two universal principles of scientific knowledge, that is (T_C) and (T_M), which belonged to two different modules.
- From these principles we inferred two theory-specific rules of research, namely, (R_{ST}) and (R_{GB}), respectively.
- On the basis of these findings we obtained metascientific knowledge of *why* certain instances of objectscientific knowledge have the conceptual representation they have.
- Consequently, we were able to account for the differences with respect to the *content* of the two theories. One possible explanation of these differences is that their *conceptual* structure contains, among other things, a *motivational* element, namely, the value of a parameter which is rooted in the motivational module.

Thus, as an immediate consequence of the hypothesis (IPS), we have *solved* our initial problem $(P1)_{ch8}$ in this way:

> $(SP1)_{ch8}$ If the metascientific extension of the generalized modularity hypothesis (MMH′) were accepted, the contribution of social factors to the conceptual structure of generative linguistic explanations would consist in the value of a parameter.

8.4 Summary

The aim of this chapter has been to show the following.

- A modular approach to the cognitive science of science is, at least to a certain extent, capable of accounting for the interaction of the 'cognitive' and 'social' factors underlying scientific knowledge in the sense of $(SP2)_{ch8}$.
- Consequently, by showing the plausibility of $(SP1)_{ch8}$ the seemingly antagonistic nature of the relationship between cognitive and sociological approaches to scientific knowledge can be resolved.
- The idea of 'linguistics self-applied' is a fruitful one in that it can be turned into analyses yielding metascientific accounts of instances of objectscientific knowledge as exemplified by the above case study on generative grammar.

The limits which these considerations are confronted with will be discussed in Section 14.3.

CHAPTER 9

Case study: A sociological extension of the holistic approach

9.1 Introduction

9.1.1 The problems

In Chapters 4 and 5 we saw, among other things, that both holistic and modular cognitive semantics may be applied to the investigation of the structure of theoretical terms. However, we concluded that the two approaches may be extended differently and that they may serve different purposes. Accordingly, Chapter 8 was an immediate continuation of the case study in Chapter 5. It exemplified an extreme case of an extension of the modular approach which led, among other things, to two main findings. Firstly, starting from the modularity hypothesis, the argumentation gave a possible account of the differences between two types of grammatical explanation. Secondly, it related some of the conceptual aspects of theory formation in generative linguistics to social ones.

In turn, we may assume on the basis of what was concluded in Chapter 6 that the cognitive theory of metaphor has perspectives very different from those of the modular approach exemplified in Chapter 8. Let us consider, therefore, another extreme case which will serve as a plausible example of possible extensions of Lakoff and Johnson's approach.

Thus we raise $(P1)_{ch9}$ which is analogous to problem $(P1)_{ch8}$ tackled in Section 8.1.1 but differs from the latter with respect to the specific theories to be analyzed:

$(P1)_{ch9}$ What kind of relationship would there be between certain 'conceptual' and 'social' factors of theories of AIDS, if the metascientific extension of the main hypothesis of the cognitive theory of metaphor were accepted?

$(P1)_{ch9}$, as opposed to $(P1)_{ch8}$, focuses on AIDS research instead of generative linguistic theories. The first reason for this is that the cognitive theory of metaphor has already been applied to the terminology of theories of AIDS. Therefore, it is only natural that we should review and evaluate these applications. As a consequence, in the present chapter, in contrast to the approach we took in Chapter 8, we will not strive to elaborate an 'original' solution to the problem at issue, but rather, our main task will be to present an evaluation of research already carried out.

A second reason why in Chapter 8 $(P1)_{ch8}$ was restricted to generative linguistic theories is that the modular framework seemed to be compatible with Bloor's Strong Programme for the sociology of knowledge right at the outset. In the debate which was provoked by the latter approach, Chomsky's generative linguistics served as one of the paradigmatic examples. Therefore, the decision that the case study should be based on generative linguistics seemed to be well motivated in the context of Chapter 8. No such immediate motivation presents itself with respect to the holistic approach.

Nevertheless, by analogy with the line of argumentation we followed in Chapter 8, $(P1)_{ch9}$ can be solved only if $(P2)_{ch9}$ is raised and answered:[1]

$(P2)_{ch9}$ Can Lakoff and Johnson's cognitive theory of metaphor as an example of holistic cognitive semantics be extended so that it captures the relationship between certain 'conceptual' and 'social' factors of scientific knowledge?

9.1.2 Ethnomethodological conversation analysis

Of course, there are many ways to discuss $(P2)_{ch9}$. Our non-trivial decision is this: we will discuss $(P2)_{ch9}$ by extending the holistic approach towards *ethnomethodological conversation analysis*. The reasons for this decision are as follows.

Firstly, ethnomethodological conversation analysis is a sociological approach to linguistic interaction which current research has tried to integrate with Lakoff and Johnson's cognitive theory of metaphor.

Secondly, an integrated approach to scientific knowledge using both the cognitive theory of metaphor and the methods of ethnomethodology has been successfully applied to the analysis of the nature of scientific theory formation in general and terminology in particular.

Thirdly, this means that ethnomethodology and certain approaches to cognitive science are not necessarily incompatible with one another at the outset. Let us illustrate the problem of their compatibility by a quotation:

> Ethnomethodological studies do not share a unified approach to research. Some ethnomethodologists, for example, reject cognitive science [...] So construed, cognitive science runs counter to the ethnomethodological project [...]
> By contrast, other researchers find that studies in cognitive sociology, cognitive psychology, cognitive anthropology, and ethnomethodology contribute to cognitive science by developing an understanding of the relation between interaction, sense-making processes, and social organization. (Saferstein 1999: 391–392)

Fourthly, applying one version of ethnomethodology to the analysis of scientific knowledge is by no means an unusual idea. In fact, one of the most influential approaches to the sociology of scientific knowledge as put forward for example in Latour and Woolgar (1979) rests on an ethnomethodological basis.[2]

Now, so as to clarify the background of the considerations to follow, some main tenets of conversation analysis need to be summarized:[3]

- Conversation analysis is rooted in ethnomethodology as elaborated, among others, by Garfinkel, Goffman, Sacks etc. It introduced new methods into linguistics. It radically rejects structuralism and poststructuralism in the generative tradition. It rejects the assumption that knowledge of language has to be conceived of as an abstract system and maintains the opposite hypothesis. According to the latter, *knowledge of language is constituted in everyday verbal interaction.*
- Consequently, it does not aim at the development of abstract theories of language but rather, the empirical and data-oriented analysis of *verbal interaction.*
- The oral data which have to be analysed are conversations *recorded* on tape.
- The recorded data are transcribed using a certain *transcription system.*
- The researcher is only allowed to analyze *natural* conversations which he or she must not influence.
- The structures analysed are the result of the *joint* communicative activity of the speakers.
- The linguistic structures work on the '*surface*' of the interaction.
- The interactive structures have to be represented with respect to their *sequentiality* and *processuality.*
- There is *order* in interaction.
- This systematic order leads to the *solution* of structural problems of interaction.

In what follows, a case study will serve to illustrate the possible solutions to $(P2)_{ch9}$ and $(P1)_{ch9}$ along the lines of the background thus sketched.

9.2 The metaphorical structure of AIDS research

9.2.1 The war scenario of AIDS research

Wolf-Andreas Liebert carried out detailed and thorough investigations with respect to the language of AIDS-research (Liebert 1995, 1996a, 1996b, 1997a, 1997b). First of all, in Liebert (1995) the conclusion reached was that talking about AIDS is fundamentally metaphorical both in expert and in popular conversation. He started from the assumption that AIDS research is governed by the metaphorical concept HIV INFECTION IS WAR which can be interpreted best by the following quotation:

> [...] the complex SCENARIO-ICM of WAR is mapped onto the SCENARIO-ICM of the HIV INFECTION. The mapping includes the projection of the parts of the whole war scenario which has typical PARTICIPANTS (COMBATANTS) with typical ROLES (AGGRESSOR, AGGRESSED, and DEFENDER) who have STRATEGIES in order to reach their GOALS (WIN THE WAR) and which furthermore has the typical stages of a BATTLE with a typical BEGINNING and END (WIN / LOSE THE WAR). Thus the HIV becomes the AGGRESSOR, the T4-CELLS become the AGGRESSED and the HUMAN PROTECTION SYSTEM becomes the DEFENDER. (Liebert 1995: 443–4)

However, complex investigations which were carried out later on did not confirm this assumption. A rich corpus of conversations which were recorded and analyzed along the lines of *conversation analysis* showed that the war-scenario played only a marginal role in the conceptualization of AIDS; rather, it was TRANSPORT, COMMUNICATION and PRODUCTION models that seemed to be dominant (see Liebert 1996a, 1997a,b). The data thus analyzed were then interpreted in the light of the cognitive theory of metaphor.

The corpus Liebert analyzed includes dialogues which differ from each other in their distance from the highly specialized theoretical language of virology. At one end of the scale there are dialogues whose participants were researchers themselves and who, accordingly, made use of the special terminology of scientific communication. At the other end of the scale, there are popular dialogues whose aim was to make scientific findings understandable to a wide audience using mass media. However different these dialogues are in other respects, they have one important common feature: although they make use of different metaphorical expressions, the metaphorical concepts behind these expressions, irrespective of the type of dialogue at issue, are the same. Consequently, scientific and everyday communication are basically governed by the same metaphorical patterns. So, scientific and non-scientific communication seem to rest on the same mechanism of metaphorical conceptualization.

This conclusion is of great importance, for it differs considerably from what the analytical philosophy of science maintains. In particular, the latter accepts the view that scientific inquiry assumes a designated position in following certain norms of rationality which do not apply, or at least only apply in a very restricted way, to everyday thinking, everyday behaviour, and everyday communication. In opposition to this, the conclusion that scientific and everyday communication rest on the same metaphorical patterns which seem to govern the conceptualization of the world suggests that scientific cognition can be embedded into the whole of human behaviour and that we are not justified in considering the former an extraordinary way of gaining knowledge of the world. Therefore, if we consider the cognitive theory of metaphor to yield, among other things, a metascientific theory capable of capturing certain aspects of scientific concept formation, then such a metatheory makes scientific inquiry appear to be significantly different from what the analytical philosophy of science assumes.

9.2.2 Conversation analysis and the cognitive theory of metaphor

The expansion of Liebert's results which led, among other things, to the conclusion that it is not only the war metaphor that is decisive with respect to the conceptualization of AIDS, was based on a corpus of conversations in German which were transcribed and analyzed with the methods of conversation analysis.[4] The corpus consisted of four components (Liebert 1997a, 1997b). Firstly, 10 authentic work meetings of virologists; secondly, a guided discussion of a group of 20 virologists about the meaning of 'science'; thirdly, interviews with 12 virologists; and fourthly, three sessions of reflexion on metaphor models with 6 virologists. After transcription the corpora were evaluated along the lines of Lakoff and Johnson (1980a) and Lakoff (1987) in order to reveal the metaphorical structure of scientific knowledge.

The main question to be answered was what results would be attained when scientists reflect on their own metaphors. So as to illustrate the way the analyses were carried out, let us quote one of Liebert's case studies in full (Liebert 1997a: 162 f.; all emphases in the original):[5]

> *Metaphor models previously introduced*
>
> DIFFERENTIATION OF HUMAN CELLS IS LEARNING IN SCHOOL (school model)
>
> *Transcript*
>
> S11 *du kannst eigentlich nur durch diese */durch dieses modifikation und instruktions(grundlagen) DURCHlaufen du kannst entweder richtig oder FALSCH ablaufen * und wenn du dann die richtige äh information und*

*kodierung mitbringst * dann wirst du in die gruppe integriert werden wenn du sie nicht mitbringst dann wirst du in den MÜLLeimer wandern oder (--) sonst was wird passieren*

S7 *oder*
sitzenbleiben und später kommen

S11 *oder*
sitzenbleiben und später kommen

Translation

S11 you can actually only go THROUGH these modification and instruction frameworks and you can either turn out right or WRONG * and if you ahm bring with you the correct information and encodings * then you will be integrated into the group if you don't bring it you will be thrown into the GARBAGE bin or (--) something else will happen

S7 or
you will repeat a grade and be late

S11 or you will repeat a grade and be late

Analysis

In previous turns the metaphor model DIFFERENTIATION OF HUMAN CELLS IS LEARNING IN SCHOOL with metaphorical concepts like SCHOOL, TEACHER, ELITE, HIGHSCHOOL DIPLOMA (ABITUR), EXAMINATION, and INSTRUCTION had already been *introduced* and discussed.

In the first turn of the transcript S11 *takes up* the metaphor model DI7ERENTIATION OF HUMAN CELLS IS LEARNING IN SCHOOL. Then S11 *adapts* the school model by *drawing conclusions in the target domain:* he states that the following proposals must be true in the target domain: 'a protein is either correct or incorrect'. This proposition could not be covered by the *introduced* school model. He *introduces* the new metaphorical model 'garbage bin' for the incorrect proteins, implicitly *introducing* the new source domain EXACT PRODUCTION which includes the model of garbage removal. Then he *adapts* it by *superimposing* it onto the school model thus creating a 'chimerical concept': The school as imagined by S11 seems to be a kind of automatic machine. The student has to run through this automatic instruction machine which encodes him/her either as right or wrong. Students with correct codes become integrated into the group, while wrong-coded ones are not students at all, they are garbage. Thus S11 expresses the perspective of the cell as a fully deterministic system, whereas the simple school model would imply at least some degree of autonomy.

S7 then *ignores* the metaphor model of the 'garbage bin' and hence the source domain EXACT PRODUCTION/GARBAGE REMOVAL *introduced* in the previous turn by S11. He *takes up* the school model and *expands* it through *adding* two *new metaphorical concepts* 'repeating a grade' and 'being late' for incorrect proteins. In showing that the school model is able to include the proposition 'a protein is either correct or incorrect' as stated by S11, S7 demonstrates that the school model is still appropriate as is the autonomy-perspective for the cell. S11 confirms this by echoing the phrase uttered by S7 ('or you will repeat a grade and be late').

If we observe this case study from the point of view of $(P1)_{ch9}$ and $(P2)_{ch9}$, unexpected findings seem to present themselves.

On the one hand, the view that metaphors play an important role in scientific inquiry is widely accepted.[6] But on the other hand there has so far been no evidence concerning the assumption that metaphors may be used *consciously* in scientific problem solving procedures, in formulating new hypotheses or in putting forward new and original ideas. Liebert's investigations thus illustrated are, however, quite remarkable from this perspective. In particular, the scope of his investigations into the metaphorical structure of the language of AIDS research was not just restricted to revealing certain metaphorical expressions and certain metaphorical concepts, but included another aspect as well:

> [...] if we gave researchers access to their own cognitive models, *making them conscious of the metaphors they live by*, then we could perhaps *stimulate* research processes, helping them play within and outside their shared metaphors. (Liebert 1995: 439; emphasis added).

This attempt to use the results of metascientific inquiry for the *improvement* of objectscientific inquiry yields the following assumption:[7]

> There is sufficient evidence from cognitive semantics and research in analogical reasoning that the reflection and changing of metaphors *can help in scientific problem-solving*. For the implementation of complex ICMs as discussed above, a new lexicographic approach, cognitive lexicography, is needed. [...] *Cognitive lexicography seen as applied cognitive semantics may therefore produce mental tools for creative problem-solving in science.* (Liebert 1995: 445; emphasis added).

Liebert *does not* explicate the metascientific consequences of his observations. Nevertheless, it is not too difficult to realize that his standpoint implies another very important difference between the metascientific application of the cognitive theory of metaphor and traditional metascience, for the latter maintains the autonomy of metascientific reflexion with respect to objectscientific research,[8] whereas the former holds that objectscientific research may be influenced *constructively* by the results of metascientific reflexion.[9]

9.2.3 Conclusions

On the basis of what has been said it seems to be the case that Liebert, when applying the cognitive theory of metaphor to the metascientific investigation of the language of AIDS research, implicitly makes use of a research strategy consisting of three steps. These three steps can be explicated as follows.[10]

Step one: Lakoff and Johnson's theory was *extended* to the *metascientific* level and so a metascientific approach was developed.

Step two: This metascientific theory was *applied* to capturing the metaphorical nature of AIDS research: both the network of metaphorical expressions and the metaphorical concepts underlying the latter were revealed.

Step three: The results of such an application were used to *improve* object-scientific research conducted by AIDS researchers. That is, the results of *meta*scientific reflexion contributed to the success of *object*scientific problem solving. The former may be used to *construct* new empirical hypotheses, new theoretical concepts and new solutions to problems within the objectscientific enterprise of AIDS research.

This research strategy is of great relevance for it shows in what way the cognitive theory of metaphor constitutes a powerful approach to the cognitive science of science whose mechanism is entirely different from that of the analytical philosophy of science.

9.3 Summary

The aim of the above considerations was to present a possible solution to $(P1)_{ch9}$ and $(P2)_{ch9}$, respectively. In sum, what has been shown boils down to the following.

- As has been illustrated by Liebert's investigations, the content of theories, the nature of empirical hypotheses and the structure of theoretical terms are shaped by metaphorical concepts which are assumed to be elements of human communication and knowledge of the world.
- One important finding, then, is the generalization of the claim already illustrated in Chapter 4 with respect to generative linguistics according to which the cognitive theory of metaphor may be used as an approach to the cognitive science of science.
- Accordingly, the approach thus characterized *widens the scope of linguistics* to a considerable extent: its object of investigation comprises not only language (in whatever sense) but certain problems of metascience as well.
- This metascientific extension rests on a research strategy consisting of three steps.

- The cognitive theory of metaphor as an approach to the cognitive science of science differs from the analytical philosophy of science in at least two respects. First, as opposed to the latter, it does not detach scientific concept formation from everyday cognitive behaviour; it rather embeds the former into the whole of human cognitive behaviour. Second, in contrast to the analytical philosophy of science, the cognitive theory of metaphor considers metascientific reflexion to be *constructive* with respect to objectscientific research in that it contributes directly to the improvement of the latter.
- These findings were, among other things, due to the fact that the empirical potential of *conversation analysis* was utilized, in that a large and rich corpus was established and carefully evaluated along the lines of conversation analysis. Conversation analysis provided the empirical basis for the metascientific use of the cognitive theory of metaphor. Thus we obtained *the solution to* $(P2)_{ch9}$:

  $(SP2)_{ch9}$ Lakoff and Johnson's approach as an example of holistic cognitive semantics is capable of being extended so that it captures the relationship between certain 'conceptual' and 'social' factors of scientific knowledge, because it can be supplemented by the methods of conversation analysis which are rooted in ethnomethodology.

- The investigations have shown that conversation analysis and holistic cognitive semantics may interact with each other so as to yield the solution of highly complicated problems which the foundations of scientific inquiry raise.[11] Thus, $(SP2)_{ch9}$ led to the following solution to $(P1)_{ch9}$:

  $(SP1)_{ch9}$ If the metascientific extension of the main hypothesis of the cognitive theory of metaphor were accepted, then theories of AIDS would be structured by metaphorical concepts constructed in the course of verbal interaction.

However plausible the case study which led to this finding may seem, its relevance is limited. We will return to this in Chapter 14.

CHAPTER 10

Conclusions: Prospects

10.1 Introduction

By way of summary, Chapters 4, 5, 8 and 9 suggest the following prospects of the metascientific application of the two approaches at issue:

(1) a. Both approaches are capable of accounting for certain central problems discussed in the philosophy of science such as
- the structure of theoretical terms, and
- the relationship between some 'conceptual' and 'social' aspects of scientific knowledge.

b. However, the considerations put forward within the two frameworks suggest very different solutions to these problems and therefore result in two incompatible accounts of certain basic aspects of scientific knowledge.

After having obtained this result and having seen how the two approaches work *if* they are applied to scientific knowledge, it is appropriate to generalize their further consequences systematically (see also the concluding paragraphs of Chapter 6). Thus the problem we have to raise in the present summarizing chapter is this:

$(P)_{ch10}$ By what further prospects can the findings of Chapters 4, 5, 8 and 9 be supplemented?

10.2 The prospects of a modular approach to the cognitive science of science

10.2.1 The double-facedness of scientific knowledge

The analytic philosophy of science, which was predominant for many decades, presupposed the unity of science. It held the view that scientific knowledge is rational, and that rationality is universal and rests on a priori criteria. Therefore, the main task of metascientific reflexion was thought to be the justification of scientific theories (see on this (13c) in Section 1.3.2). This meant that the structure of scientific theories was related to the assumed universal and a priori criteria of rationality. On the basis of such a comparison the results of scientific research were judged as rational or irrational, true or false, successful or unsuccessful etc. This activity was a kind of 'superscientific' enterprise which was governed, among other things, by two commitments (Quine uses the term 'suprascientific'). According to the first, metascientific reflexion is entitled to select the a priori criteria of scientific rationality which are used to evaluate theories. Secondly, in the course of these judgements metascience makes use of methods which differ fundamentally from those of its object of investigation, i.e. objectscientific research.

While the scope of the analytic philosophy of science thus characterized was restricted primarily to the analysis of universal properties of scientific knowledge, and the specific aspects of particular disciplines and theories were not focussed on, certain recent developments have chosen the opposite task. To mention just one radical example, the Strong Programme for the sociology of knowledge, as elaborated by Bloor and his Edinburgh School, strives to replace the analytic philosophy of science by arguing, among other things, for the assumption that rationality is neither universal nor a priori, but rather, that there are different rationalities rooted in social factors (Bloor 1976, 1983, 1997, Barnes et al. 1996).

It is easy to realize that both views are one-sided: neither of them asks the question as to the *relationship* between the *general*, universal aspects of human behaviour and those properties which are *specific* to certain scientific communities, to particular scientific disciplines or to individual theories. Therefore, we could construct a plausible argument for the workability of a modular approach to the cognitive science of science if it were possible to show that it accounts for this relationship.

Such an argument may rest on the assumptions introduced in Section 2.2.2. It will lead, among other things, to two complementary conclusions. Firstly, from the metascientific extension of the generalized modularity hypothesis (MMH′) and the assumption that modules include universal principles (see (2) in Section 2.2.1), it follows that, just as in the case of other domains of behaviour, there are universal principles that underlie scientific knowledge. Secondly, from (MMH′) and the fact that instances of behaviour are based on the interaction of different modules (see (2) in Section 2.2.1), it follows that instances of scientific knowledge are based on the interaction of modules.

Thus, a modular metatheory that presupposes (MMH′) is capable of capturing the essential *double-facedness of scientific knowledge*: firstly, scientific knowledge is based on the interaction of *universal* principles, and secondly, the interaction of these universal principles yields the *specific* aspects of particular instances of scientific knowledge which are not to be reduced to instances of other domains of behaviour.

10.2.2 The problem of the uniqueness of scientific knowledge

The analytic philosophy of science attributes a unique position to scientific knowledge in so far as the latter is — in opposition to everyday behaviour and everyday knowledge — assumed to meet the standards of rationality. However, recent approaches which intend to transgress the limits of the analytic philosophy of science — like Bloor's aforementioned radical Strong Programme for the sociology of knowledge — trace scientific knowledge back to the whole of human behaviour and try to understand it on the basis of everyday knowledge. In this respect, a modular metatheory may yield interesting findings as well. In particular, it can capture the relationship between scientific knowledge and everyday cognitive behaviour in a compelling way, because the metascientific extension of the generalized modularity hypothesis (MMH′) as introduced in Section 2.2.1 leads to at least two further consequences. Firstly, it follows that scientific and non-scientific (everyday) knowledge rest on the interaction of the *same* universal principles belonging to the *same* modules; therefore, scientific knowledge is to be *integrated* into the whole of human cognitive behaviour. Secondly, those features of scientific knowledge which distinguish it from other fields of behaviour can be inferred from the parametrization relation between universal principles belonging to different modules. Accordingly, it is possible to obtain those specific properties which *distinguish* scientific knowledge from other areas of cognitive behaviour.

10.2.3 The explication of the term 'theory'

Along the lines of (MMH′), scientific knowledge — as part of human behaviour — is based on the interaction of the same universal principles which other fields also rest on; thus *scientific theories must be regarded as instances of behaviour which rest on the interaction of universal principles* (see also note 8 in Chapter 8). Consequently, it will be possible to capture the structure of scientific theories as a set of representations in the *same* way as other instances of behaviour are captured. This means that the structure of theories can be described in terms of conceptual, grammatical, motivational etc. representations (see (i) in (1) in Section 2.2.1 and (MMH1′) in Section 8.2.1). As a result, a modular metatheory may achieve the description/explanation of scientific theories as 'conceptual structures', 'social constructs', 'conventions', 'syntactic structures', etc. within a unified framework that accounts for this wealth of different aspects. At the same time this means that a modular approach to the cognitive science of science may *integrate* those metatheoretical tools that can capture this richness of aspects and may, in the sense of Chapter 8, work as an integrated approach to the cognitive science of science (this is in full accordance with (MMH′) and (17) in Section 2.2.1.). Such an approach goes far beyond the facilities of other metascientific approaches and suggests a novel explication of the term 'theory', which may be worth considering. In fact, it is capable of accounting for the richness of approaches which Laudan, in arguing against Bloor's overemphasis on sociological aspects, put in this way.

> [...] The fact of the matter is that science is a multifaced process. One could well say that science is a psychological phenomenon (considering, for instance, the role of cognition and perception in it) and thus should be studied primarily by psychologists. [...] Alternatively, science is a goal-directed activity and is thus legitimately in the sphere of decision theory and operations research. Insofar as science is carried out out by human animals it is presumably a biological activity. The point is that science can be legitimately studied in a variety of ways. (Laudan 1984: 66–67).

10.2.4 Self-application

One interesting peculiarity of a metascientific application of modular cognitive semantics is its self-applicability: the fact that we may reflect on the nature of theory formation in linguistics by making use of certain methods used in linguistics itself.[1] Therefore, linguistics seems to be a reflexive discipline which, under certain conditions, may be used to reveal its own properties by the metatheoretical application of its own methods.

10.2.5 The pragmatics of scientific discourse

The pragmatics of scientific discourse[2] is a typical inter- and transdisciplinary field of research characterized, among other things, by the fact that traditional disciplinary boundaries are both widened and transgressed with respect to research concerning certain important aspects of scientific knowledge which have been neglected by the analytic philosophy of science.

In this context the term 'pragmatics' is understood in a wide sense and indicates the fact that the investigation of scientific knowledge must not be restricted to its logical, syntactic and semantic components; rather, research should, besides these, substantially focus on its communicative, textual and social factors as well. Therefore, anyone who is interested in the pragmatics of scientific discourse has to face the complexity of the problems involved. Even if one strives to restrict the subject matter of one's research to well-defined properties of precisely determined corpora based on oral conversations or written texts, owing to this complexity any kind of investigation will have to touch on a complicated network of at least epistemological, communicative, conceptual, semiotic, textual, ontological and further factors.

Therefore, the questions that have to be asked are these:

(2) How can,
  a. the complex nature of the problems raised by the *object* of investigation, and
  b. the *methodological* problems raised by the transgression of conventional disciplinary boundaries

  be handled?

A possible modular approach to the cognitive science of science gives clear answers to these questions:

- According to (11a) in Chapter 1, one of the main strategies of modularism is the subdivision of the object of investigation into subsystems which can be captured by the (mainly formal) means at our disposal. Therefore, the complex domain of 'the pragmatics of scientific discourse' has to be replaced by a set of distinct but interacting modules. Consequently, as was emphasized in Chapter 4, the term 'pragmatics' is to be discarded; rather, it is the specific interaction of modules that is expected to account for the problems which have conventionally been labelled 'pragmatic' (in whatever sense). Thus, as regards (2b), it is the replacement of the domain 'pragmatics' by certain relatively autonomous subsystems that yields new boundaries between the fields of investigation.
- In (i) in (1) in Section 4.2.1 we enumerated the modules which are generally assumed within the framework of the two-level approach. In Chapter 5 we have seen that two of these, namely, the grammatical and the conceptual — which are connected by an interface called semantic representation —,

are responsible for some relevant properties of theoretical terms. Moreover, in Chapter 8 it was shown that further properties of scientific concept formation and scientific explanations can be captured if we focus on the interaction between the motivational module on the one hand and the grammatical and conceptual ones on the other hand. Accordingly, as for (2a) above, the problems concerning the object of investigation are related to these modules and reformulated as problems concerning the specific interaction between them.

- Consequently, what has been revealed in previous chapters within the framework of the two-level approach can be interpreted as a modular account of the highly complex problems raised by the pre-explicative notion of 'pragmatics of scientific discourse'.
- Basically, this approach consists of two steps of argumentation. Firstly, in accordance with the 'thesis of simplification' the object of investigation (i.e. the 'pragmatics of scientific discourse') is subdivided into certain modules (relatively autonomous subsystems such as the grammatical, the conceptual, and the motivational module). In the second step the particular interaction of these modules is revealed via the parametrization relation between the universal principles which the modules comprize.

Nevertheless, this is not everything that can be achieved with respect to the modular treatment of the phenomena at issue. For example, in Bierwisch (1980) a theory of social interactions was outlined (cf. the module of social interactions in (i) in (1) and in (7) in Section 2.2.1) whose implications have not been touched on in our study so far. Following this line of research, further important contributions to the 'pragmatics of scientific discourse' can be expected, and there is a wide range of possibilities to be considered.

10.3 The prospects of a holistic approach to the cognitive science of science

10.3.1 The double-facedness of scientific knowledge

The cognitive theory of metaphor explains the nature of theoretical terms and certain aspects of theories on the basis of the assumed metaphorical structure of conceptualization. It may account for the question of the relationship between the *general, universal aspects* of conceptualization and those properties which are *specific* to certain scientific communities, to particular scientific disciplines or to individual theories — although in a very different way from the modular approach (as outlined in Section 9.2.1). The crucial problem in this respect is one of the central quandaries of the cognitive theory of metaphor concerning the identification and scope of metaphorical concepts.[3] On the basis of Lakoff and Johnson (1980a), (1999) we are justified in assuming that on the one hand there are metaphorical concepts whose validity is general; and that on the other hand, there must also exist metaphors which are specific to certain types of conceptualization. Thus in principle the cognitive theory of *metaphor may account for the double-facedness of scientific knowledge.*

10.3.2 The problem of the uniqueness of scientific knowledge

While the analytic philosophy of science attributes a unique position to scientific knowledge, recent approaches try to account for it on the basis of everyday knowledge. In particular, the cognitive theory of metaphor interprets complex and abstract concepts in terms of concrete, elementary experiences and this is not unique to everyday concept formation but can be extended to scientific theories as well. Consequently, according to Baldauf (1997: 278) one may assume a continuum of conceptual domains which comprises both the abstract concepts of everyday knowledge and those of scientific knowledge as extensions of the former. Scientific knowledge is to be integrated into the whole of human cognitive behaviour:

> A Scientific Theory attempts to provide an understanding of some class of phenomena through the consistent elaboration of some set of metaphorical concepts. When *the basic metaphors of a scientific theory are extensions of basic metaphors in our everyday conceptual system,* then we feel that such a theory is 'intuitive' or 'natural'. (Lakoff and Johnson 1980b: 207; emphasis added)

10.3.3 The explication of the term 'theory'

In close connection with the last point, we may achieve a new explication of the notion of 'scientific theory' as emphasized in Baldauf (1997: 279). As we know, in the course of scientific research one attempts to establish a conceptual framework which captures conceptual domains not accessible to immediate experience. In this process, metaphors transform a set of individual data into a coherent whole and thus yield a hypothesis. One precondition for putting forward a hypothesis in this way is the correspondence between the source domain and the target domain. The second precondition is that, in establishing such a correspondence, certain methodological rules such as those of consistency or coherence have to be followed. If these conditions are met, "the metaphor selected constitues a theory" (Baldauf 1997: 279). Consequently, theories may be defined in terms of the metaphorical concepts which rest on a given correspondence between the source domain and the target domain.

10.3.4 Self-application

i. *The meta-metascientific extension of the cognitive theory of metaphor.* As a first example of the self-applicability of the cognitive theory of metaphor, let us review the investigations carried out by O. Jäkel with respect to the metaphorical structure of philosophies of science (Jäkel 1996, 1997). Jäkel analysed the language of six outstanding philosophers of science who played a decisive role in our thinking on the nature of scientific inquiry. First of all, Jäkel showed that these six philosophers of science, however different their views of science are, seem to assume a generalized scenario whose components are the following.

- *scientist*: agent, engaged in scientific study
- *nature*: the scientist's object of study
- *method*: the scientist's activity, approach to the object
- *theories*: result of scientific activity
- *scientific progress*: decision/choice between competing theories (Jäkel 1996: 649)

Jäkel showed that the metaphorical expressions which the philosophers of science use rest on metaphorical concepts connected to the above scenario.

Aristotle: SCIENCE IS VISION
Descartes: SCIENCE IS A JOURNEY
Bacon: SCIENCE IS THE COERCION OF NATURE
Kant: SCIENCE IS THE BUILDING OF AN EDIFICE
Popper: SCIENCE IS AN ARMED STRUGGLE FOR THE SURVIVAL OF THE FITTEST THEORY
Kuhn: SCIENCE IS A RELIGIOUS WAR OVER THE ACCEPTANCE OF A CERTAIN KIND OF GAME

The metaphorical scenarios of science investigated with respect to these metaphorical concepts can be summarized in Table 10.1 (Jäkel 1996: 671; 1997: 276).

Table 10.1

	SCIENCE	scientist	nature	method	theories	scientific progress
Aristotle	VISION	passive spectator, observer	objects of observation	–	–	
Descartes	JOURNEY	traveller	–	straight path, slow but steady movement	–	forward and upward movement
Bacon	COERCION OF NATURE	offender, inquisitor, torturer, intruder	female, hiding valuables; victim of coercion	coercion and dissection of nature	–	snatching more valuables from nature's innards
Kant	BUILDING AN EDIFICE	pioneer, architect, construction worker	–	inspection of the development site, architectural design of the building	buildings	completion of the edifice
Popper	STRUGGLE FOR SURVIVAL	warrior	–	weaponry	rivals competing in the struggle for survival	elimination of weaker theories evolutionary selection of the fittest
Kuhn	RELIGIOUS WAR & GAME	fanatical believer, serviceman, puzzle-solver	toy items to be forced into boxes	rules of the puzzle-solving game	charis-matic religious leaders	revolution-ary change of leadership, victory in a religious war

We may draw the following conclusions from Jäkel's investigations thus summarized.

- The conceptualization of science, as witnessed by the work of outstanding philosophers of science, is structured by metaphorical concepts.
- During the history of the philosophy of science the focus of these metaphors was different (see (27) in Section 2.3.1 on the focussing function of metaphors).

The important conclusion that the cognitive theory of metaphor can be applied on the *meta*-metascientific level thus allowing conventional boundaries of linguistics to be radically transgressed, is expressed by Jäkel in the following way.

> Once more, Cognitive Linguistics proves useful not only for an understanding of the nature of language, but also for an understanding of *other* human cognitive achievements, in this case for a better understanding of the essence of different philosophical approaches to the nature of *science*. (Jäkel 1996: 675; emphasis added)

ii. *The metaphorical structure of theories of metaphor.* As a second example, let us consider the argument in (3).

(3) *If*
  a. according to the cognitive theory of metaphor abstract conceptual domains are structured metaphorically, and
  b. theories of metaphor themselves constitute abstract conceptual domains,

then
  c. theories of metaphor themselves are structured by metaphorical concepts.

(3c) implies, on the one hand, that the language of theories of metaphor consists of metaphorical expressions, on the other hand that the occurrence of the latter is not individual or casual, but rather, is the manifestation of metaphorical concepts structuring scientific knowledge conducted within theories of metaphor. Jäkel (1997: 115 ff.) discusses a couple of examples which show that the conceptualization of metaphors in the course of theory formation in linguistics is itself metaphorical. For example, right at the outset the etymology of the term *metaphor* reflects the metaphorical concept METAPHOR IS TRANSPORT. Another example may be the idea that we can understand the nature of metaphor if we see it as a person: METAPHORS ARE PERSONS; let it be sufficient here to refer to expressions like *dead metaphor* or *living metaphor* (Richards 1971, Ricoeur 1986).

More important than these examples is the self-applicability of the cognitive theory of metaphor. For example, when Lakoff and Johnson speak of MAPPING between the source domain and the target domain, they make use of a concept which goes back to mathematics but which in mathematics is also used metaphorically (Jackendoff and Aaron 1991: 335, Jäkel 1997: 116). Or, alternatively, the metaphorical concept METAPHORS ARE PICTURES seems to be manifested in Lakoff and Johnson's term of *projection* (cf. Jäkel 1997: 116).

However superficial these examples are, they show that theories of metaphor are typical abstract domains of cognition, and that they belong to theory formation in cognitive semantics. This means that the cognitive theory of metaphor can be applied to the metascientific investigation of cognitive semantics itself. That is, the cognitive theory of metaphor yields the self-applicability of holistic cognitive semantics.

Consequently, in this sense the cognitive theory of metaphor is a specific manifestation of the cognitivist interpretation of the Quinean program: *metascience may be cognitive semantics self-applied.*

10.3.5 The pragmatics of scientific discourse

As we mentioned in Section 10.2.5, the main task of the modular approach to the cognitive science of science is to subdivide the complexity of problems in order to decide which can be captured within a certain subdomain by the use of certain methods and which not. In contrast to this, 'the thesis of complexity' presupposed by holism (see (10a) in Chapter 1) suggests that for those who try to understand the pragmatics of scientific discourse by means of the cognitive theory of metaphor the question is basically not how to exclude certain aspects and to involve others. Rather, one has to ask:

(4) Which aspects of the complex nature of the object of investigation should be emphasized more strongly and which should be pushed into the background while still acknowledging their presence?

What we have shown in the previous chapters is, with respect to this question, the following:

- In Chapter 4 it was shown that metaphorical concepts underlying theoretical terms are connected to scenarios and that, in accordance with this, they are inseparable from aspects which are conventionally (in a pre-explicative sense) regarded as 'pragmatic'.
- In Chapter 9 we saw, nevertheless, that one may go one step further by integrating the cognitive theory of metaphor with conversation analysis. As is well known, the roots of conversation analysis lie not only in the sociological theory of ethnomethodology, but also in 'pragmatics'.
- Consequently, the way the cognitive theory of metaphor is applied to the investigation of scientific knowledge is one possibility of treating complex problems which touch on 'the pragmatics of scientific discourse'.

In this respect, our considerations in the previous chapters seem to pave the way for future research concerning the 'pragmatics of scientific discourse'. Primarily, such research should focus on the relationship between metaphorical conceptualization and social interaction as manifested in conversations. Methodologically, the possible integration of conversation analysis and the cognitive theory of metaphor should be further refined.

10.4 Summary

The answer to $(P)_{ch10}$ is as follows.

$(SP)_{ch10}$ a. The *prospects* of the cognitive theory of metaphor as an example of holistic cognitive semantics and of the two-level approach as an example of modular cognitive semantics result from the fact that they radically *widen* the scope of linguistic research. Among other things, they are capable of reformulating and solving classic problems and current quandaries in the philosophy of science.

b. Although the premises of the modular and the holistic approach are very different, both can be extended in a way which enables them to capture analogous problems of the philosophy of science. The findings they yield are original, fruitful and radically different from the perspectives of the analytic philosophy of science. In particular, both approaches are capable of accounting for central metascientific problems within their particular framework such as
   - the structure of theoretical terms,
   - the relationship between certain 'conceptual' and 'social' aspects of scientific knowledge,
   - the double-facedness of scientific knowledge,
   - the problem of the uniqueness of scientific knowledge,
   - the explication of the term 'theory',
   - the self-applicability of cognitive semantics, and
   - the pragmatics of scientific discourse.[4]

c. However, the considerations put forward within the two frameworks suggest very different solutions to these problems. They draw two different pictures of the way scientific

> knowledge in general and scientific concept formation in particular are structured and proceed.

In parts II and III our task was to illustrate certain *prospects* of modular and holistic cognitive semantics *if* they are applied to an extreme case of cognition, namely, scientific knowledge. $(SP)_{ch10}$ summarizes the findings we have obtained so far and yields *a partial solution to (Q''')*. In the next part of the book we will *focus on certain potential limits* of the metascientific applications of the two approaches at issue.

Part IV

Limits

Chapter 11

The background

11.1 Introduction

Just as in Parts II and III, so too in the last part of this work we will start from crucially important and familiar problems of the philosophy of science. We will select three which typically reveal some of the limits of metascientific reflexion.

The first problem to be raised is one of the most fundamental issues of the naturalized philosophy of science: namely, the question of whether the abandonment of the analytic philosophy of science also means the abandonment of the normativity of metascientific reflexion.

The second problem concerns the role which demonstrative inferences play in the philosophy of science. The demonstrativeness of inference is one of the major requirements which the rational reconstruction of scientific theories in the sense of the analytic philosophy of science has to fulfill. Therefore, the question arises whether cognitive semantic approaches to the cognitive science of science meet this requirement, too.

Finally, as is well known, the emergence of fallacies counts traditionally as an obvious limit of epistemological thought and the analytical philosophy of science. Thus the question is whether approaches to the cognitive science of science are confronted with these limits.

The next three sections will illuminate the relevance of these quandaries. As usual, the discussion will be as simple as possible and instead of giving a detailed overview of the state of the art our remarks will be confined to merely emphasizing those aspects of the problems which seem to be relevant for understanding the argumentation in the case studies.

11.2 Constructivity

The method which the analytic philosophy of science makes use of is called *rational reconstruction.* The method of rational reconstruction basically includes two components. Rational reconstruction is *descriptive* in that it analyzes the structure of existing scientific theories. At the same time, however, rational reconstruction has a *normative* component as well: its central task is not revealing the structure of scientific theories as it is, but rather, of the way it ought to be if it satisfied the *a priori* principles of rationality. See the following classic summary of the relationship between the normativity and the descriptivity of the analytic philosophy of science:

> *Is the philosophy of science descriptive or normative?* [...] As is very often the case in rational discussions, here too it is misleading to think in exclusive alternatives. However, such thinking cannot be ruled out by answering the question with a vague and uncompelling 'as well as'. Rather, one has to show to what extent the *rational reconstruction* of scientific knowledge, which is the subject matter of the philosophy of science, possesses both a *descriptive* and a *normative component.* Revealing this in a precise and detailed manner is possible only with respect to specific and detailed investigations within the philosophy of science. (Stegmüller 1978: 8; emphasis in the original; my translation, A. K.)

The normativity of rational reconstruction is, however, extremely *limited* in so far as the analytic philosophy of science is not expected to contribute to the practical solution of the objectscientific problems which the everyday activity of working scholars focuses on.

Now, one of the central points of discussion is the assumption that naturalized philosophy of science intends to dispense with normativity completely (see e.g. Goldman 1986, Kim 1988). This assumption rests on a straightforward argument: if, following Quine (1969a), epistemology and the philosophy of science should proceed as science does, then they ought to become descriptive and explanative, abandoning normativity. However, this interpretation is an oversimplification and needs refinement. Not only current views, but also Quine's later writings admit that naturalism does not exclude normativity at the outset:

> Naturalism of epistemology doesn't jettison the normative and settle for the indiscriminate description of ongoing procedures. For me normative epistemology is a branch of engineering. It is the technology of truth-seeking, or, in a more cautiously epistemological term, prediction [...] There is no question here of ultimate value, as in morals; it is a matter of efficacy for an ulterior end, truth or prediction. The normative here, as elsewhere in engineering, becomes descriptive when the terminal parameter is expressed. (Quine 1986: 664–65).

Nevertheless, it is far from clear what the source of normativity is, and how the normativity of naturalized epistemology / the naturalized philosophy of science works in particular cases. Therefore, there has been a wide discussion of these issues in the literature and many different conceptions have been suggested.[1] Without going into a systematic overview of the literature, it will be sufficient to quote a possible point of departure:

> Apart from the desire for an ultimate foundation for knowledge, a common objection to naturalistic approaches in the philosophy of science is that such approaches are limited to the mere description of what scientists do. The philosophy of science, it is claimed, is concerned to develop normative models of how science should be pursued — that is, models of scientific rationality. But there is a way in which naturalists can also be normative [...]. Naturalists are not limited merely to describing what scientists say and do. They can also develop theoretical explanations — for example, cognitive or evolutionary explanations — of how science works. Just as theoretical mechanics provides a basis for designing rockets, so *a powerful theoretical account of how science works could provide a basis for sound advice on scientific practice and policies.* (Giere 2001b: 310; emphasis by underlining in the original, italics added)

This is, indeed, *very different* from the normativity of the analytic philosophy of science. In fact, it is this kind of normativity which was exemplified in Chapter 9 with respect to the application of the cognitive theory of metaphor to the language of AIDS research and which we called there *constructivity*.

The term 'constructivity' will be retained here as well so as to refer to that specific kind of normativity which was characterized in the above quotation (the problem of the normativity of naturalized epistemology in the sense of constructivity is extensively elaborated on in Kertész 2002b, 2004; see also Finke 1982 for this term). By 'constructivity' we mean *the capability of a metascientific approach to contribute to the improvement of objectscientific research.*

Now, the fact that the present study focuses not only on the prospects but also on the *limits* of cognitive semantic approaches to scientific knowledge boils down to the following question:

(1) What are the limits of the constructivity of the naturalized philosophy of science?

It is worth pointing out the relevance of the question in (1). In particular, it raises a really important issue which may substantially reveal the use and scope of naturalized philosophy of science. Should it be possible to show that the metascientific reflexion on object-scientific research substantially contributes to the solution of objectscientific problems, then the usability of such metascientific reflexion would clearly go far beyond the limits of the analytic philosophy of science. In this case metascience would be capable of *influencing* the content of scientific knowledge as well as the progress of science, and scholars would be well-advised to make use of this capacity in consciously integrating it into their problem-solving activity. Therefore, from the point of view of the evaluation of the prospects and limits of any version of naturalized philosophy of science it is a crucial question *how far* this constructivity can go and *where those limits lie which it cannot transgress.*

11.3 Plausible reasoning

One of the basic assumptions of the analytic philosophy of science is that scientific inquiry proceeds rationally. Although within the analytic philosophy of science there are different views concerning the question of what rationality is, there seems to be agreement on the fact that one thing that rationality involves is that the theses of a theory must be connected by valid logical inference. To illustrate this fundamentally important requirement, let us quote Stegmüller again who in one of the volumes of his monumental series on the classic methods of the analytic philosophy of science enumerates 9 criteria of the acceptability of scientific theories. The very first one is this: "(1) […] Are the logical deductions correct?" (Stegmüller 1970: 373). The requirement that statements should be systematized in a logically demonstrative manner applies not only to scientific theories, but to the analytic philosophy of science *itself*: its main tool is formal logic and it uses a logical inventory so as to 'rationally reconstruct' the structure of scientific theories. However, recent research into the nature of scientific inquiry has shown that although strict logical validity may be a reasonable requirement, in many cases it is only one of the ideals which traditional philosophy of science maintains but which cannot be realized in practice. Thereby the following considerations seem to be relevant.

Logically valid inferences are also called *demonstrative* (Polya 1948, 1954). Nevertheless, the non-demonstrativeness of an inference does not mean automatically that it is incorrect. Although *plausible inferences* are deductively not valid, they are frequently used ways of reasoning both in everyday argumentation and in scientific inquiry. The main property of plausible inferences is that they lead to conclusions which are not necessarily true, but merely count as credible to a certain extent. (The classic literature on plausible inference includes Polya 1948, 1954, Rescher 1976, 1987 etc.)

As a first approximation, Polya characterizes plausible reasoning by pointing out that it differs in at least two relevant aspects from demonstrative inferences:

First, they do not have the certainty of a strict demonstration. Second, they are useful in acquiring essentially new knowledge, and even *indispensable to any not purely mathematical or logical knowledge*, to any knowledge concerned with the physical world. We would call the reasoning that underlies this kind of evidence 'heuristic reasoning' or 'inductive reasoning' or (if we wish to avoid streching the meaning of existing terms) '*plausible reasoning*'. *We accept here the last term.* (Polya 1948: 221–222; emphasis added)

Demonstrative	*Heuristic*
If *A* then *B*	If *A* then *B*
B false	*B* true
----------------	----------------
A false	*A* more plausible

The comparison of these patterns may be instructive. It may grant us an insight, not easily obtainable elsewhere, into the nature of plausible (heuristic, inductive) reasoning.

Both patterns have the same first premise:

If *A* then *B*.

They differ in the second premise. The statements:

B false B true

are exactly opposite to each other but they are of similar 'logical nature', they are on the same 'logical level'. The great difference arises after the premises. The conclusions

A false A more credible

are on different logical levels and their relations to their respective premises are of a different logical nature.

The conclusion of the demonstrative syllogism is of the same logical nature as the premises. Moreover, this conclusion is fully expressed and is fully supported by the premises. If my neighbour and I agree to accept the premises, we cannot reasonably disagree about accepting the conclusion, however different our tastes or other convictions may be.

The conclusion of the heuristic syllogism differs from the premises in its logical nature; it is more vague, not so sharp, less fully expressed. This conclusion is comparable to a force, has direction and magnitude. It pushes us in a certain direction: *A* becomes *more* credible. The conclusion has also a certain strength: *A* may become *much more* credible, or *just a little more* credible. The conclusion is not fully expressed and is not fully supported by the premises. *The direction is expressed and is implied by the premises, the magnitude is not.* For any person, the premises involve that A becomes more credible (certainly not less credible). Yet my neighbour and I can honestly disagree *how much* more credible *A* becomes, since our temperaments, our background, and unstated reasons my be different. (Polya 1948: 222–223; italics in the original)

In mathematics and the natural sciences which make use of formal methods there is a clear difference between the problem-solving mechanisms and the systematization of the results of such mechanisms. Whereas the process of problem solving involves mainly plausible reasoning, the theses which are put forward as the result of problem solving have to be proved in a logically demonstrative way.

In many other disciplines, however, the exposition of the results of research also reflects the way the corresponding problems were solved and remains strongly argumentative in nature. In this case very often even the final form of presenting the results makes use of plausible rather than demonstrative inferences. It goes without saying that at least the two paradigmatic examples of cognitive semantic theories whose possible metascientific extensions we are discussing in this book belong to the latter group. Certainly, the cognitive theory of metaphor explicitly rejects not only formalization but also the emphasis on logical validity that is one of the major cornerstones of the analytic philosophy of science (see (30) in Section 2.3.1). The two-level approach is very cautious in evaluating its own results; its proponents often emphasize the preliminary and heuristic nature of their assumptions and the inferences they draw from their basic hypotheses (see e.g. Bierwisch and Lang 1989: 445).

Now, if the cognitive theory of metaphor and the two-level approach are metascientifically extended, then those approaches to the philosophy of science which they manifest will carry the features of the original objectscientific theories. If the latter make use of plausible reasoning rather than of demonstrative inferences, then this will apply to their metascientific use as well. Therefore, the question is this:

(2) What are the consequences of the use of plausible inferences in the metascientific extensions of the two-level approach and the cognitive theory of metaphor?

Depending on what consequences can be revealed, the possibility must not be excluded that with respect to the use of patterns of inference the cognitive science of science may be considerably more *limited* than the analytic philosophy of science. Should this be the case, far-reaching consequences concerning the reliability of cognitive scientific approaches to scientific knowledge would follow. Therefore, it is well motivated to examine whether this is a *real limit* of cognitive semantics and if so, what its consequences for the metascientific extensions of cognitive semantics are.

11.4 Fallacies

Fallacies are erroneous arguments which violate the accepted standards of correct inference. Some general properties of fallacies are these.

- The classic fallacies are informal, because to decide whether a fallacy has been committed or not the context of argumentation has to be taken into acount. The logical structure of the argument itself is not enough to make such a decision. Not every instance of logically invalid inference is fallacious:

  > An argument that is correct or appropriate in one context of dialogue may be incorrect or fallacious in another context. [...] An argument may have a fallacious form, but not every instance of an invalid form of inference is an invalid inference. The reason is that a given argument can have many forms, some valid, and some invalid. If an argument is an instance of an invalid form of inference, it does not follow that the argument must be invalid. Whether the argument is fallacious or not depends on whether and how that form of inference has been used in the *context* of a dialogue. (Walton 1997: 213)

  For example, certain non-demonstrative inferences may be plausible and not fallacious — as we have seen in the previous section. It is important to emphasize that *very often it is only the context thay decides whether a certain inference is a fallacy or a plausible argument.*[2] What primarily matters is effectivity: whereas plausible inferences are effective tools of gaining new information, fallacies are ineffective.
- Although fallacies have been recognized and discussed by early Greek philosophers and although they played an important role in philosophy and the sciences, their structure has only recently been made subject to systematic reflexion.[3]
- Although, partly as a result of their context-dependence, there are no general criteria to draw a strict dividing line between non-demonstrative fallacious inferences and non-demonstrative correct inferences such as plausible arguments, there is a classic list of informal fallacies which have been known from ancient times on.[4]

One of the reasons why fallacies deserve special attention is that in traditional epistemology and the analytic philosophy of science invalid arguments whose fallaciousness has been explicitly shown often indicate important *limits*. For example, the central aim of analytic epistemology is to *justify* knowledge of the world in general, and the central aim of the analytic philosophy of science is to justify scientific knowledge in particular (see (13c) in Section 1.3.2.). Now, for example, arguments which intend to justify knowledge but which cannot avoid the fallacies of infinite regress or circularity may indicate that the scope of epistemology and/or the philosophy of science is limited, because they cannot reach one of their main aims, namely, the justification of knowledge.

However simplified this account is, it seems to motivate the following question.

(3) Can the metascientific application of cognitive semantic approaches avoid fallacies and their consequences?

The relevance of the question in (3) is this: if it can be answered affirmatively, then with respect to fallacies cognitive semantic approaches to the cognitive science of science seem to transgress obvious *limits* of traditional epistemology and/or the analytic philosophy of science. If the answer is negative, then of course, these limits are retained and the scope of cognitive semantic approaches to the cognitive science of science must be treated with caution.

11.5 Summary

In the last part of the present book certain possible limits of cognitive semantic approaches to the cognitive science of science will be discussed. Just as in parts II and III, we chose certain central and widely disputed problems of the philosophy of science as our point of departure. With respect to the problem of the constructivity of naturalized philosophy of science, in Chapter 12 a case study will be developed which centers around question (1), but it will also touch on question (3). The two case studies in Chapter 13 are devoted to (2) and (3), but in the course of the argumentation they will also reveal the relationship between the latter and question (1). Finally, Chapter 14 will summarize some of the substantial limits of the two approaches at issue.

Chapter 12

Case study: The sceptical dilemma of cognitive semantics

12.1 Introduction

In the fifties and sixties, the principles of the analytic philosophy of science influenced the methodological presuppositions and objectives of generative grammar. In (13) in Chapter 1 we outlined a very simplified summary of some basic tenets of the analytic philosophy of science. The main tenets said that the latter is a *philosophical* discipline, it presupposes certain *a priori* assumptions concerning the nature of the criteria of rationality, and it strives to *justify* the results of scientific inquiry in so far as, on the basis of the principles of rationality, it evaluates scientific knowledge as true or false, rational or irrational, justified or not.

In the seventies and early eighties these principles were exposed to a heated debate in theoretical linguistics. The debate centred on the question of whether contemporary theoretical linguistics — which was established by and which developed dynamically in connection with the birth and success of generative grammar — fulfills its own methodological requirements in that it is an empirical and explanatory discipline, or whether, on the contrary, it departs from its own desiderata by being in fact a non-empirical, hermeneutic enterprise.[1] However, the debate led to destructive consequences. Firstly, because the antagonism between the two positions could not be resolved and neither position was therefore able to handle the issues raised. Secondly, because the working linguist had good reason to conclude that metascientific reflexion would never support him/her in tackling and solving his/her objectscientific tasks; accordingly, metascientific considerations were, from his/her point of view, useless. The attempt to evalute linguistic inquiry along the lines of the analytical philosophy of science, as summarized in (13) in Chapter 1, led thus to the following conclusion.

(T) The analytic philosophy of science
   a. is not compatible with the practice of linguistic inquiry, and

b. cannot contribute to the solution of objectscientifc problems which the linguist deals with.

This pessimistic view seems to have been presupposed implicitly in most theoretical works in current linguistic research, and, not infrequently, it is even expressed explicitly. The following quotation from one of the most influential works of the past two decades may illustrate this judgement:

> There is *a wide gap* between general methodological principles on the one hand, and on the other hand the specific procedural, analytical, descriptive, and theoretical decisions one faces in conducting *the daily affairs of a particular field.* (Langacker 1987: 33; emphasis added)

However, parallel with the debate, new and decisive changes occured: the naturalistic turn and the cognitive turn. As a result of these changes the current state of the art both in linguistics and the philosophy of science differs considerably from those circumstances which motivated (T). Therefore, the question must be asked *whether the relationship between the practice of linguistic inquiry and metascientific reflexion can be reconsidered today.* Accordingly, the following problem arises.

$(P)_{ch12}$ Can the cognitive science of science contribute to the solution of the objectscientific tasks of cognitive semantics?

This is the problem which the case study to be presented focuses on. On the basis of this case study we will argue for the following hypothesis which is the antithesis to (T) and which is intended to be the solution to $(P)_{ch12}$:

$(SP)_{ch12}$ a. The cognitive science of science and the practice of cognitive semantics are trivially *compatible* with each other.

b. Metascientific reflexion on the nature of cognitive semantics carried out by the methods of the cognitive science of science may *contribute immediately* to the solution of the object-scientific tasks of cognitive semantics.

c. At the same time the scope of this contribution is very *restricted* and metascience must not take over the tasks of object-scientific inquiry.

In the case study to be developed so as to support this hypothesis we will focus on the relationship between holistic and modular cognitive semantics.

12.2 The metaphorical circle and the modularistic circle

As a starting point, for convenience let us repeat the main hypothesis of the cognitive theory of metaphor and the generalized modularity hypothesis already introduced in Sections 2.2.1 and 2.3.1, respectively:

(1) *The main hypothesis of the cognitive theory of metaphor (HH'):* Human cognitive behaviour is structured metaphorically.

(2) *The generalized modularity hypothesis (MH')*: Human cognitive behaviour is organized in a modular way.

Now, if (1) and (2) are given, the linguist has to face the following question:

(3) Should (1) or (2) be accepted?[2]

Unfortunately, however, it is not easy to answer this question. One reason for the difficulty is, of course, that both (1) and (2) are intended to be 'empirical' hypotheses which one can choose between only on the basis of 'empirical' data; but at present the data which would facilitate an empirically well-founded decision are not available (see for this e.g. Reis 1987, Bierwisch and Lang 1989).[3] However, there is another, even more destructive, reason. In particular, both theses result in an fallacy: (1) yields the *metaphorical circle* and (2) the *modularistic circle*. Let us consider the metaphorical circle first.

Any application of the cognitive theory of metaphor is endangered by circularity, because of the argumentative relationship between metaphorical expressions and metaphorical concepts documented in (20a) and (b) in Section 2.3.1. As witnessed by the following two quotations, the cognitive theory of metaphor "is a clear case of circular reasoning" (McGlone 2001: 95):

> Metaphors as linguistic expressions are possible precisely because there are *metaphors in a person's conceptual system* (Lakoff and Johnson 1980a: 6; emphasis added).

> Since metaphorical expressions in our language are tied to metaphorical concepts in a systematic way, *we can use metaphorical linguistic expressions to study the nature of metaphorical concepts* and to gain an understanding of the metaphorical structure of our activities. (Lakoff and Johnson 1980a: 7; emphasis added).

Accordingly, we obtain the following.

(4) *The metaphorical circle*:
[...] Lakoff's claim that metaphors transcend their linguistic manifestations to influence conceptual structure rests solely on these manifestations. How do we

> know that people think of theories in terms of buildings? Because people often talk about theories using building-related expressions. Why do people often talk about theories using building-related expressions? Because people think about theories in terms of buildings. Clearly, the conceptual metaphor view must go beyond circular reasoning of this sort and seek evidence that is independent of the linguistic evidence. (McGlone 2001: 95)

That is, the holistic stance leads to an *untenable situation*. Nevertheless, modularism is no less problematic, either:

(5) *The modularistic circle*:
   a. One has to put forward the *empirical hypothesis* that knowledge of language is based on the interaction of relatively autonomous systems, *because*, for *methodological reasons*, the object of investigation can be grasped only by systematic simplifications, namely, its subdivison into separate (sub)systems (see (11a, b)) in Section 1.2.3.
   b. At the same time, however, we have to follow *the methodological principle* that the object of investigation has to be subdivided into relatively autonomous systems, *because* we have accepted the *empirical hypothesis*, according to which knowledge of language consists of subsystems.

It is easy to grasp the circularity of modularistic argumentation if we remind the reader of one of the quotations, which may illustrate the modularity hypothesis, and which formulates the basic *empirical* assumption of modular cognitive semantics:

> All human behaviour is essentially organized in a modular fashion. The structure formation that underlies a behaviour instance *V* is the joint product of relatively autonomous, functionally interacting systems and subsystems. (Lang 1989: 266).

(5) can be understood immediately if we compare this account of the modularity hypothesis with the following definitely *methodological* reasoning from the work just quoted:

> In view of their role in the structure formation underlying a behaviour instance *V*, two systems S_i and S_j are *relatively autonomous* to the extent that
> (1) S_i and S_j are determined each by specific principles, and hence,
> (2) a theory capable of accounting for *V* as the outcome of two separate systems S_i and S_j is superior to a theory that construes *V* as a product of a uniform system *S*. (Lang 1989: 268)

In the same volume, in summarizing their findings, Bierwisch and Lang repeat the whole of the circular argumentation as follows:

> Singling out autonomous subsystems of mental organization and specifying their content *is based on the assumption that the structure and the functioning of cognitive systems do indeed have such a modular structure.* Autonomous subsystems and structural levels of the theory, from this factual point of view, correspond to autonomous subsystems in the actual structure of mental states and processes. This actual structure, however, cannot be discovered directly but only in the form of explanatory theories. From this theoretical pont of view, identifying subsystems and explaining their autonomy, as well as finding out the nature of their interaction, becomes an aspect of adequate theory formation. *Autonomous components of the theory are then justified (a) by the independence of the basic concepts, (b) by the internal structure of the theoretical components which the basic concepts enter and (c) by the possibility of formulating regularities in the framework of (a) and (b) which explain the facts on the appropriate level of abstraction.* (Bierwisch and Lang 1989: 495; emphasis added)

It is interesting to remark that the modularistic circle may be detected also in Chomsky's writings. Starting from his remarks concerning Chomsky's approach, Müller (1991) relates the modularistic circle to cognitive science in general:

> It is an *incorrect logical inference* which is found again and again in modularistic theories of cognitive science to interpret restrictions motivated by research strategic considerations as realistic claims about the organization of mind and brain. That fact that perhaps we can obtain knowledge of the world or of our own mind only through a modularistic 'theory filter' does not mean that the world and the mind are modular. (Müller 1991: 407; my translation, A. K.; emphasis added)

That is, the proponents of modularism support the application of a methodological principle with an empirical hypothesis and the empirical hypothesis with the application of a methodological principle.

Consequently, *the problem of the linguist* is even more difficult than (3) suggests. The task is no longer simply to decide whether one should accept (1) or (2), but rather, to choose between *two fallacies*:

(6) Should the linguist choose the modularistic circle or the metaphorical circle?

This problem has two important aspects:

- First of all, (6) is a *dilemma*, because it commits one to a choice between two alternatives which are equally destructive.
- Secondly, this dilemma is of a *sceptical* nature, because it questions the possibility of scientific knowledge within the domain of cognitive semantics.

Therefore, (6) will be called *the sceptical dilemma of cognitive semantics.*

12.3 The sceptical dilemma

What course can the cognitive linguist take now, in the light of the sceptical dilemma? Basically, there appear to be at least three possibilities:

(7) a. The first is, of course, *to give up* cognitive semantics as a scientific enterprise.
b. As another attempt, one may try to *solve* the dilemma by showing that at least one of the two fallacies which led to the dilemma was obtained by incorrect argumentation and therefore does not occur.
c. The third possibility is simply to *ignore* the dilemma. This means that, although both fallacies are acknowledged, the working linguist considers them to be irrelevant for his/her daily research and continues his/her activity as if the dilemma did not exist.

Now, can (7a) be maintained? Since Kuhn (1970) it is commonplace that scientific theories cannot be discarded on the basis of internal anomalies or fallacies or contradictions or errors of whatever kind; scientific theories can only be replaced by a new theory which is in certain respects more attractive and more acceptable for a particular scientific community. Therefore, the solution indicated by (7a) is, for well-known metascientific reasons, not possible right at the outset.

(7b) would mean that, since the sceptical dilemma is a typical philosophical problem, it may be resolved by philosophical methods. However, Quine's seminal work (1969a) showed that philosophical argumentation can never be successful against scepticism. Therefore, *philosophical* reflexion on the nature of scientific knowledge has to be replaced by the application of *empirical* science in order to solve or to eliminate those problems which have traditionally been the subject matter of epistemology and the analytical philosophy of science. Consequently, due to the basic tenets of the cognitive science of science, the philosophical resolution of the sceptical dilemma is also not available at the outset.

What remains is (7c), which presupposes an essential distinction between the philosophical nature of the sceptical dilemma and the 'empirical' tasks of the working linguist. The latter may assume that he or she, as an empirical researcher, must not raise philosophical problems and that philosophical issues are clearly outside his/her interests. The working linguist may simply ignore the existence of the sceptical dilemma and conduct his/her research as if there were no basic philosophical problems concerning the possibility of scientific knowledge. Let us call this stance *the strategy of ignorance.*

The strategy of ignorance seems to work very well, at least in natural science. The problem of induction is a good example of its workability. Induction is the central method of natural science according to which from past experience we draw inferences concerning the future behaviour of objects. Hume raised the famous problem of whether induction can be justified; that is if we can reasonably draw inferences from past experience which go beyond this experience and concern future properties of objects. His answer was *no*: "*we have no reason to draw any inference concerning any object beyond those of which we have had experience*" (Hume 1739, bk 1, pt III, sec. XII; emphasis original). Thus the method of induction cannot be justified and there arise sceptical doubts. Although since Hume's work there is no denying that the inductive method cannot be justified and that there are no philosophical tools at our disposal which would be effective against the sceptical doubts concerning induction, nevertheless, despite its unjustifiability, induction is the basic method of natural science: in everyday practice it turns out to be workable, although we are aware of its limits. No natural scientist would think of giving up his/her scientific aims by referring to the unsolvability of Hume's problem. (See on this also Strawson 1985.)

Analogously, we may assume that the sceptical dilemma of cognitive semantics can be handled following the same pattern as the problem of induction; that is, by making use of the strategy of ignorance. However, this conclusion is *incorrect*. It is incorrect, because it is not compatible with the basic assumption of cognitivism which — as we have seen in Section 1.2.3 in (2c) and (3c) — says that the primary aim of the subdisciplines of cognitive science in general and of cognitive linguistics in particular, is to capture *classic philosophical problems about the nature of mind and cognition with empirical methods*. Surely, cognitive linguistics need not solve *all* philosophical problems; but there are *certain* philosophical problems which it simply *must* try to solve so as to avoid a self-contradictory stance. In particular, the sceptical dilemma as a philosophical problem is rooted in two empirical hypotheses concerning the structure of cognition — i.e. the holistic hypothesis and the modularity hypothesis. Which of these hypotheses is acceptable and which not is without doubt one of the central questions of cognitive science and cognitive linguistics. Therefore, we must not claim that the philosophical problem that is associated with the relationship between these two hypotheses is not interesting to the cognitive linguist and can, therefore, be ignored — unless we are ready to admit that our view is contradictory. In sum, the attempt to apply the strategy of ignorance to the sceptical dilemma would be (due to the specific nature of the linguistic subdiscipline of cognitive science as indicated in (3) in Section 1.2.1) *self-contradictory*. Therefore, *(7c) must be rejected* as well.

In this way all the possibilities enumerated in (7) are ruled out. It seems to be the case that the sceptical dilemma cannot be captured. However, this is a devastating consequence, because as long as the dilemma exists, cognitive semantics as a scientific research programme seems to be seriously questionable.

But cognitive semantics need not, of course, be given up for the above reasons. We only have to realize that the aims of cognitive semantics are compatible with those of the cognitive science of science which is rooted in naturalized epistemology. This finding implies that we have to solve the sceptical dilemma by using *the 'empirical' methods of the cognitive science of science itself*. How can such a solution be found, and what would it amount to?

12.4 The solution to the dilemma

First of all, a brief summary of the line of argumention carried out so far is in order:

a. As a starting point, we reflected on cognitive linguistic inquiry *metascientifically* in that we focussed on the relation between its empirical hypotheses and methodological principles.
b. As a result of this metascientific reflexion, we detected the sceptical dilemma of cognitive semantics which corresponds to a *classic philosophical problem*.
c. We saw that this problem (i) could not be solved by philosophical methods, (ii) nor could it be ignored, and (iii) we saw that cognitive semantics cannot be simply given up, either.
d. As a last chance we guessed that the way out may be to reflect on the empirical hypotheses which motivated the cognitive dilemma by using the empirical methods of cognitive semantics itself. This empirical self-reflexivity is a *typical* strategy of naturalism, which Quine puts in this way: "Epistemology, for me, is only science self-applied." (Quine 1969b: 293)

This brief survey may imply the following argumentative steps which are expected to lead to the avoidance of the dilemma:

e. One thing that self-reflexion involves is revealing *the reason why* the sceptical dilemma occurred.

f. If we ask this question, it immediately turns out that (1) and (2) *overgeneralize.* For example, the generalized modularity hypothesis (MH') as put forward in (2) claims that "*all* human behaviour is essentially organized in a modular fashion." (Lang 1989: 266; emphasis added).

With respect to its generality, the main hypothesis of the cognitive theory of metaphor is characterized by McGlone in this way.

> While Lakoff stresses the role of metaphor in conceptual representation, neither he nor his colleagues have offered a detailed model of how metaphoric representations are constructed or used. In the absence of an explicit model from proponents of the conceptual metaphor view, Murphy (1996, 1997) formulated two versions of what such a model might look like, a strong version and a weak version. In the strong version, *all* concepts other than those based directly on sensorial-perceptual experience have no intrinsic structure of their own. Instead, they are represented entirely as a set of mappings from the representational structure of more concrete concepts. (McGlone 2001: 93; emphasis added).

Therefore, as an answer to the question mentioned in (e), let us hypothesize that *the reason why the sceptical dilemma occurs is that (1) and (2) overgeneralize.* From this assumption it follows that the dilemma may be avoided if the overgeneralization is ruled out.

g. Let us see whether it is possible to avoid the overgeneralization. Indeed, (1) and (2) do not apply to the same data, because certain data may be explained on the basis of (1), and others with respect to (2) — the case studies in parts II and III of this book have witnessed this quite convincingly (see for this Kiefer 1995 in another context). Therefore, it seems to be well-motivated to weaken both hypotheses. Nevertheless, this finding is *not new* but has already been documented in the relevant literature. Let it be sufficient to illustrate this by just two plausible examples.

In his paper cited above McGlone concludes with respect to the cognitive theory of metaphor:

> While the strong version of the metaphoric representation claim is *theoretically untenable*, Murphy's (1996) weak version might be *more plausible.* According to this version, abstract concepts are not exclusively piggybacked on concrete concepts but are nonetheless influenced by their conceptual structure. Metaphor still plays a role in organizing the abstract concept, but the representation of the abstract concept is not representational per se. [...] The weak version thus assumes that metaphor plays a causal role in the structure of abstract concepts but is not the sine qua non of their conceptual representation.
>
> In contrast to the strong version, the weak version of the metaphoric representation claim is open to empirical investigation. A reasonable test of the

> claim would, at a minimum, involve three steps. First, one would identify an abstract concept for which the set of idiomatic expressions in a particular culture suggests a conceptual metaphor, such as THEORIES ARE BUILDINGS. Next, one would explore the idomatic expressions used in another culture to describe the concept and determine whether this culture employs a different metaphor. Third, having established that members of the different cultures talk about theories in different ways, one would then seek to demonstrate that they think about theories in different ways, as evidenced by performance in nonlinguistic reasoning about theories. This third step is crucial, for without it there is no empirical basis for the claim that conceptual metaphors transcend their linguistic manifestations [...] (McGlone 2001: 94–95)[4]

The reason why the metaphorical circle emerges is that, in accordance with the strong version

> at present, however, conceptual metaphor research has not yet gone beyond the first step of the investigation. Lakoff and his colleagues base the metaphoric representation claim solely on intuitions about how certain idioms thematically cohere. (McGlone 2001: 95.)

Consequently, if instead of the strong version the weak version were practiced, the circle could be avoided.

The second example is Harder's notion of 'partial autonomy' which witnesses that the same result can be achieved by a less overgeneralizing account of modularism. This notion is much more differentiated than the notion of autonomy which the two-level approach or Fodor's or Pylyshin's modularism presuppose, "because partial autonomy places the difference of levels within the context of an overall continuity" (Harder 1999: 197). Moreover, Harder emphasizes:

> Partial autonomy is a difficult concept, but one that is essential to come to terms with, not only in linguistics and cognitive science, but in scientific description generally. The concept should not be understood merely as a wishy-washy compromise solution, giving both parties a little of what they want, but as a hard fact about the way the world works. (Harder 1999: 198)

Should this more sophisticated account be accepted, the modularistic circle would not occur. (There are several other attempts to avoid the overgeneralization of the two approaches mentioned; see, for example Gibbs 1996)

h. Moreover, if both the metaphorical circle and the modularistic circle can be avoided, then the sceptical dilemma would not occur, either. So, although the dilemma could not be solved, new conditions were construed on the basis of which the sceptical dilemma need not be present. Accordingly, the original philosophical problem has been *eliminated.* This outcome of the argumentation was due to the empirical self-reflexion of cognitive semantics; and this self-reflexion, in turn, corresponds to what one expects from naturalism.

i. How could this result be achieved? It could be achieved, because the line of argumentation put forward in a.–h. is nothing but the manifestation of a procedure basically consisting of the following three phases:
   - Firstly: we reflected metascientifically on two incompatible hypotheses of cognitive semantics and in this way we confronted a *classic philosophical problem*, namely, the sceptical dilemma of cognitive semantics (see a.–c.).
   - Secondly: We tried to capture this problem by the self-reflexion of an empirical discipline, namely, that of cognitive semantics itself, along the lines of the principle of "science self-applied" (cf. d.–e.).
   - Thirdly: The result of this metascientific self-reflexion was used to *improve* objectscientific research by showing that the scope of both (1) and (2) have to be restricted and that (1') and (2') are more plausible. Thus, *meta*scientific reflexion contributed to the solution of an *object*-scientific task, namely, the possible modification of empirical hypotheses (see f.–h.).

12.5 Summary

The present chapter was devoted to the question of whether the cognitive science of science can contribute to the solution of the objectscientific tasks of cognitive semantics. In other words we tried to find out whether the cognitive turn justifies the maintenance of $(SP)_{ch12}$, which was contrasted with (T). The case study carried out supports $(SP)_{ch12}$ and thus provides a possible solution to $(P)_{ch12}$ in the following way.

We have seen that cognitive semantics is, due to (3c) in Chapter 1, fully compatible with the central assumptions of the cognitive science of science, although the former was developed independently of the latter. Therefore, *$(SP)_{ch12}$ (a) can be accepted* trivially. The reason the cognitive science of science may be considered to yield a workable metatheory of cognitive semantics is that the empirical hypotheses and methodological objectives of cognitive semantics are, by definition, compatible with those of the cognitivist manifestation of assumptions put forward by Quine within his naturalized epistemology.

It also turned out that precisely that type of self-reflexion which was carried out in the last section, in addition to corresponding to the basic strategy of naturalism, also implies a specific capacity of metascientific reflexion which is missing from the analytic philosophy of science and which was not recognized by Quine, either. In particular, metascientific reflexion may contribute immediately to the solution of objectscientific problems, thus it may improve cognitive semantic research and it may further the development of cognitive semantics directly. Accordingly, the plausibility of *(SP)*$_{ch12}$*(b)* was shown. This finding has a twofold relevance. Firstly, it enriches the facilities of the cognitive science of science with a *new* perspective. Secondly, (SP)$_{ch12}$(b) is a strong argument *against the working linguist's claim* quoted in Section 12.1, according to which metascientific reflexion is pointless and useless, because it does not affect the success of everyday research. In fact, (SP)$_{ch12}$(b) suggests the opposite. It admits, on the one hand, that the only thing that, from the point of view of the working linguist, legitimizes systematic metascientific reflexion is its contribution to the success of cognitive linguistic research itself; and on the other hand, it states that such a contribution is possible.

However, the above case study points out the *limits* of this stance, too. It is *very important* to emphasize that the metascientific reflexion carried out led to a seemingly *trivial* result: namely, the necessity of restricting the scope of overgeneralizing empirical hypotheses. *It is the triviality of this result that will yield a significant conclusion.*

As we have seen, the insight that (1) and (2) are too strong is not new, but has already been put forward in the literature (cf. for example McGlone 2001, Murphy 1996, 1997, Harder 1999). More precisely, this means that metascientific reflexion of the kind exemplified in this case study comprises two basic aspects. Firstly, it showed the *implausibility* of certain empirical hypotheses. Secondly, it supported the *plausibility* of certain modifications of these hypotheses which are already available in the literature.[5] Thus, although metascientific reflexion can *contribute* to the solution of objectscientific problems, it is not capable of *solving* objectscientific problems per se; that is, it is not for example capable of suggesting an entirely new empirical hypothesis which should be accepted instead of (1) and (2). Surely, the linguist must not expect more from metascientific reflexion than this, because the stronger claim that objectscientific problems should be solved on the metascientific level would evidently be untenable. Consequently, we have supported *(SP)*$_{ch12}$ *(c) as well* and, as a result, obtained *the answer to (P)*$_{ch12}$ as indicated in Section 12.1.

In the light of this outcome of our considerations it is easy to realize the peculiarities of the relationship between cognitive semantics and metascientific reflexion:

- It seems to be useful to elaborate a naturalized metatheory of cognitive semantics, because the former may contribute to the workability, the effectiveness and the success of the latter.
- Nevertheless, metascientific reflexion can never replace objectscientific research and the task of the cognitive *linguist* will always *differ* from that of the *meta*linguist.

In trying to differentiate between the philosophy of linguistics and the philosophy of language, Auroux and Kouloughli put forward a similar idea:

> Of course we do not imagine that philosophers as such could generate, by themselves, new linguistic knowledge. They have to work with linguists in creative interaction, to provide them with relevant pieces of information and suggest possible directions in which to seek solutions, to generate critical discussions, and pose precise questions. *The philosopher of linguistics does not have to work in the field of linguistics exactly like a linguist.* But as far as the philosophy of linguistics is concerned, his purposes do not basically differ from the linguist's. *The basic objective is to help generate new linguistic knowledge.* (Auroux and Kouloughli 1993: 38; emphasis added).

The case study certainly motivates the identification, systematization and application of those strategies which facilitate the objectscientific use of metascientific reflexion by revealing the potential of the latter, but without denying its limits.

Chapter 13

Two case studies: Cognitive semantics and classic philosophical problems

13.1 Introduction

In (2b) in Section 1.2.1 we claimed that one of the central tasks of cognitive science was to transform very old philosophical questions concerning the nature of the mind into empirical questions and to answer them by using the methods of empirical disciplines. Moreover, in (2a) in Section 1.2.1 we also maintained that linguistics was one of the subdisciplines of cognitive science. These two claims yielded (2c) which we repeat here for the sake of later reference as (1):

(1) One of the central aims of a certain kind of linguistics is to reformulate, to answer or to eliminate classic philosophical questions concerning the nature of the mind and knowledge by dealing with them empirically and interdisciplinarily.

It is worth illustrating the workability and the relevance of (1) by a plausible example.[1] The example will be taken from Chomsky's work because there is no denying that it played a decisive role in the development of cognitive science.[2]

In the programmatic writings which Chomsky put forward in the eighties he referred to Plato's Meno-paradox (see e.g. Chomsky 1986). The Meno-paradox is one of the most fundamental *classic* problems of philosophy which questions the possibility of human knowledge. Plato introduces the paradox as follows:

> Men. Why, on what lines will you look, Socrates, for a thing of whose nature you know nothing at all? Pray, what sort of thing, amongst those that you know not, will you treat us to as the object of your search? Or even supposing, at the best, that you hit upon it, how will you know it is the thing you did not know?
> Soc. I understand the point you would make, Meno. Do you see what a captious argument you are introducing — that, forsooth, a man cannot inquire either about what he knows or about what he does not know? *For he cannot inquire about what he knows, because he knows it, and in that case is in no need of inquiry; nor*

> *again can he inquire about what he does not know, since he does not know about what he is to inquire.* (Plato: Meno, 80d-e. In: Plato with an English Translation by W. R. M. Lamb. Cambridge, Mass.: Harvard University Press and London: William Heinemann, 1962; emphasis added)

Here evidently there is a classic philosophical problem which can be reconstructed like this:

(2) *Classic philosophical problem:* How is knowledge possible, if our empirical experiences of the world are uncertain, vague and incomplete?

Socrates answers this question by referring to a paradox: on the one hand we cannot inquire about what we know, because we already know it; on the other hand we cannot inquire about what we don't know either, because we don't know what we should inquire about. As long as this paradox is unsolved, we have to admit that knowledge is not possible. How can the paradox be solved? Plato's classic philosophical solution is this:

> Soc. [...] Seeing then that the soul is immortal and has been born many times, and has beheld all things both in this world and in the nether realms, she has acquired knowledge of all and everything; so that it is no wonder that she should be able to recollect all that she knew before about virtue and other things. For as all nature is akin, and the soul has learned all things, there is no reason why we should not, by remembering but one single thing — an act which men call learning — discover everything else, if we have courage and faint not in the search; since, it would seem, *research and learning are wholly recollection.* (Op.cit. 81c-d; empasis added)

That is:

(3) *Classic philosophical solution*:
a. Our immortal soul had beheld everything before it was born; after we have been born, we remember these experiences and, therefore, what we call learning is nothing but recollection.
b. Consequently, when we are born, we already have knowledge.
c. Therefore, the paradox does not appear.
d. Therefore, knowledge is possible.

Now, exactly along the lines of (1), Chomsky tries to *reformulate* the question in (2) as a problem of empirical linguistics (see e.g. Chomsky 1986):[3]

(4) *The reformulation of the classic philosophical question as a problem of empirical linguistics:* How is it possible that children acquire their mother tongue in a very short time in spite of the fact that the information they get from their environment is insufficient, vague and uncertain?

Let this strategy be illustrated by plausible quotations. In reformulating the classic problem, Chomsky first reinterprets it with respect to 'cognitive systems':

> The essence of Plato's problem was well expressed by Bertrand Russell in his later work when he raised the question: 'How comes it that human beings, whose contacts with the world are brief and personal and limited, are nevertheless able to know as much as they do know?' [...] The problem that arises when we consider the matter with a little care is one of 'poverty of the stimulus'. Although our cognitive systems surely reflect our experience in some manner, a careful specification of the properties of these systems on one hand, and of the experience that somehow led to their formation on the other, shows that the two are separated by a considerable gap, in fact, a chasm. The problem is to account for the specificity and the richness of the cognitive systems that arise in the individual on the basis of the limited information available. Cognitive systems result from the interaction of experience and the organism's method of constructing and dealing with it, including analytic mechanisms and the intrinsic determinants of maturation and cognitive growth. The problem then is to determine the innate endowment that serves to bridge the gap between experience and knowledge attained [...] (Chomsky 1986: xxv-xxvi)

Secondly, this problem is reduced to the nature of the language faculty:

> Much of the interest of the study of language, in my opinion, lies in the fact that it offers an approach to Plato's problem in a domain that is relatively well circumscribed and open to *inspection and inquiry*, and at the same time deeply integrated in human life and thought. If we can *discover* something about the principles that enter into the construction of this particular cognitive system, *the principles of the language faculty*, we can progress *toward a solution for at least one special and quite important case of Plato's problem.* We can then ask whether these principles generalize to other cases, or if not, whether an approach that meets with a degree of *explanatory success* in the case of human language can at least serve as a suggestive model for similar inquiries in other cognitive domains. (Chomsky 1986: xxvi; emphasis added)

The second quotation shows explicitly in what way typical expressions belonging to the vocabulary of empirical scientific research (e.g. "inspection", "inquiry", "explanatory success", "discover" etc.) enter the reformulation of the classic problem.

Chomsky answers (4) in that he projects it onto the basic hypotheses and theoretical terms of his linguistic theory which he considers to be an empirical one:

(5) *Empirical solution:* The reason why children acquire their mother tongue in a very short time in spite of the fact that the information they get from

> their environment is insufficient, vague and uncertain is that the language faculty is innate.

In accordance with this solution, the last quotation goes on like this:

> My own belief is that these principles do not generalize, that they are in crucial respects *specific to the language faculty*, but that the approach may indeed be suggestive elsewhere, both in its achievements and their apparent boundaries. (Chomsky 1986: xxvi; emphasis added)

Moreover, the solution to Plato's problem can be further characterized in the following way:

> There is reason to believe that knowledge of language [...] should be regarded as a system of principles that develops in the mind by the fixing of values for certain parameters on the basis of experience, yielding systems that appear to be highly diverse but that are fundamentally alike in deeper respects. [...] the system of knowledge attained is largely *preformed*, as much *a part of our biological endowment* as is the general organization of our body. It seems that Plato's problem should be addressed along these lines. (Chomsky 1986: 272; emphasis added)

Thus we may conclude that one of the most influential linguistic theories was motivated, among other things, by the idea that classic philosophical questions concerning the nature of knowledge and mind can be reformulated as 'empirical' problems to be solved by the 'empirical' methods of linguistics.

In Chapter 1 we have seen that it is (2c) (repeated here as (1)) which generative linguistics shares with cognitive semantics. Therefore, a serious problem arises. Namely, on the one hand, it is a *fact* that the linguistic subdiscipline of cognitive science (which includes both generativism and trends called cognitive semantics) maintains (1). But on the other hand, the question may be asked if generative linguistics *correctly* maintains (1). Some contemporary metatheoreticians of linguistics give a negative answer to this question. For example:

> The ecumenical euphoria of the 1980s has evaporated, and *no one any longer believes that generative linguistics might magically solve philosophical problems about the mind* [...]. (Harré and Harris 1993: ix; emphasis added)

This quotation raises the following question:

(6) Does *cognitive semantics* correctly maintain (1)?

The point is that we cannot know at the outset whether the alleged failure of generative linguistics to proceed along the lines of (1) claimed in the last quotation is due to certain properties of the linguistic subdiscipline of cognitive science, which includes both generative linguistics and cognitive linguistics, or to those pecularities of generative linguistics which the latter does not share with cognitve linguistics.[4] Therefore, we are justified in asking to what extent cognitive semantics is still entitled to maintain (1). Thus we obtain $(P1)_{ch13}$ and $(P2)_{ch13}$:

$(P1)_{ch13}$ Could classic philosophical problems be reformulated and solved empirically, if the metascientific extension of the generalized modularity hypothesis (MMH′) were accepted?

$(P2)_{ch13}$ Could classic philosophical problems be reformulated and solved empirically, if the metascientific extension of the main hypothesis of the cognitive theory of metaphor (MHH′) were accepted?

We will proceed as follows. Firstly — as usual, within the frames of a case study — in Section 13.2 $(P1)_{ch13}$ will be reduced to a more specific problem $(P1')_{ch13}$. Secondly, in Section 13.3 $(P2)_{ch13}$ will be reduced to $(P2')_{ch13}$ and a case study presented. The conclusions which will be drawn from the case studies with respect to $(P1')_{ch13}$ and $(P2')_{ch13}$ will lead to a possible answer to $(P1)_{ch13}$ and $(P2)_{ch13}$, respectively.

13.2 Case study 1: Modularism

13.2.1 The problem of the Criterion of Truth

It is well known that in scientific argumentation case studies are especially useful if they explore the boundaries of possibilities. Let us, therefore, sharpen the thesis mentioned in (1) to *an extreme case.* (1) implies that one central task of cognitive semantics is the empirical reformulation, solution or elimination of classic philosophical questions (see Chapter 1). We have seen that Meno's paradox was a paradigmatic example of the workability of this thesis with respect to generative grammar. Accordingly, so as to test the two-level approach as a paradigmatic manifestation of modular cognitive semantics, we have to find *another problem* which

(i) is a classic philosophical problem;
(ii) which is no less significant than the Meno-paradox; and
(iii) which can be sharpened in a direction which goes beyond the implications of the Meno-paradox.

The *problem of the criterion of truth* meets these requirements. It played an important role in the debates between philosophical schools of the hellenistic age. One of the classic formulations of the problem is attributed to Sextus Empiricus. Briefly, the essence of the problem is whether there is such a criterion on the basis of which the truth or falsity of our claims can be decided. The often quoted classic lines raising this problem are the following.

> [...] in order to decide the dispute which has arisen about the criterion, we must possess an accepted criterion by which we shall be able to judge the dispute; and in order to possess an accepted criterion, the dispute about the criterion must first be decided. And when the argument thus reduces itself to a form of circular reasoning the discovery of the criterion becomes impracticable, since we do not allow them [the dogmatics] to adopt a criterion by assumption, while if they offer to judge the criterion by a criterion we force them to *a regress ad infinitum.* And furthermore, since demonstration requires a demonstrated criterion, while the criterion requires an approved demonstration, they are forced into *circular reasoning.* (Sextus Empiricus: Outlines of Pyrrhonism, Eng. tr. by R. G. Bury. Cambridge, Mass: Harvard U. P., 1961 [1933], II. 20; emphasis added)

The quotation is easier to understand if we know that proving the circularity of an argument or showing that it leads to an infinite regress belong to the five Modes of sceptical argumentation. Sextus Empiricus discusses these "Five Modes leading to suspension" (op.cit. I. 164–173), which are attributed to Agrippa, in the following way.

> The Mode based upon regress *ad infinitum* is that whereby we assert that the thing adduced as a proof of the matter proposed needs a further proof, and this again another, and so on *ad infinitum*, so that the consequence is suspension, as we possess no starting-point for our argument. [...] The mode of circular reasoning is the form used when the proof itself which ought to establish the matter of inquiry requires confirmation derived from that matter; in this case, being unable to assume either in order to establish the other, we suspend judgement about both. That every matter of inquiry admits of being brought under these Modes we shall show briefly in this way. (Sextus Empiricus op.cit. I. 166–169.)[5]

The problem of a criterion of truth (hence: the problem of the *Criterion*) raised by Sextus Empiricus in the way thus illustrated reflects the requirements mentioned in (i)–(iii) above in the following manner:

Ad (i). The problem originates in ancient scepticism and is undoubtedly one of the most serious quandaries of classic philosophy which seems to yield a strong argument for scepticism and to undermine the possibility of knowledge.[6]

Ad (ii). The problem of the Criterion manifests itself in seminal works which play a decisive role in the history of philosophy: Montaigne's Rouet, Chisholm's Criterion, Rescher's Wheel, Fries' trilemma, Albert's Münchhausen trilemma.[7] One proposal for a simplified reconstruction of the common core of these versions of the problem is the following.[8]

One may ask the question why we accept a sentence *a* as being true. The answer is: because we obtained *a* by using a certain method *m*. But why did we apply *m*? The first alternative is to say that we applied *m* because *m* led to true sentences. In this case, however, our argumentation is circular, since we supported *a* with *m* and *m* with *a*. According to the second alternative we accepted *m* as a method yielding true sentences, because there is another method m_1 which yielded *m*. But why did we apply m_1? Because m_1 can be accepted on the basis of another method m_2 — and so on ad infinitum. In sum, the reason why there is no criterion for the truth of our sentences is that the argumentation results either in circularity or in an infinite regress.

Ad (iii). This simple reconstruction also shows in what sense the problem of the Criterion corresponds to (iii) above. In particular, the manifestations of the problem of the Criterion mentioned in (ii) focus primarily on the relationship between *the truth of scientific claims* and *scientific method*. This is an important observation in so far as it suggests that if modular cognitive semantics — by the analogy of the Meno paradox — were able to offer an empirical solution to the problem of the Criterion thus reconstructed, it could unexpectedly open new prospects: it would be able, among other things, to motivate the construction of *metascientific* theories.[9] So, if modular cognitive semantics facilitated an empirical reformulation and solution of the problem of the Criterion, then its consequences would go far beyond those of Chomsky's treatment of the Meno-paradox.

Therefore, we reduce $(P1)_{ch13}$ to the following question:

$(P1')_{ch13}$ Could the classic philosophical problem of the Criterion be empirically reformulated and solved, if the metascientific extension of the generalized modularity hypothesis (MMH′) were accepted?

Let us see what kind of solution to $(P1')_{ch13}$ follows from the modularity hypothesis.

13.2.2 The argumentation strategy

The problem of the Criterion rests on a pattern of inference which leads to circularity or infinite regress. Since in this problem the structure of inference is focussed on, we have to reflect on the argumentation strategy of the possible solution as well. Therefore, in this section we will briefly survey the inferences which underlie the generalized modularity hypothesis (MH′) and the meta-scientific extension of the generalized modularity hypothesis (MMH′).

The modularity hypothesis was introduced into linguistic theories in many different ways.[10] The core of these diverse accounts which we repeat here for convenience is the following assumption:

(7) Knowledge of language is organized in a modular way.

The relevance of (7) is rooted, among other things, in the fact that it constitutes one premise of the central argument for the assumption that cognition as a whole is modularly organized. This argument rests on an inference which is a frequently used problem-solving strategy in cognitive science. This inference leads to very strong empirical hypotheses, but is, nevertheless, in most cases only applied implicitly. It is a kind of modular manifestation of the idea according to which in cognitive linguistics "one can move freely and gradually from facts about language to facts about human cognition and further on to facts about human life generally" [...] (Harder 1999: 196; also quoted in Section 1 of the Introduction). The general structure of the inference may be reconstructed like this:

(8) *If*
  a. *x* is part of *y*, and
  b. *x* has the property *P*,
  then
  c. also *y* has the property *P*.

Let us assume that

x = knowledge of language
y = cognition
P = is organized in a modular way

Then from (7) we can infer (9) on the basis of (8):

(9) Cognition is organized in a modular way.

Since (8) is one of the most typical inferences used in cognitive science, indeed, it is not restricted to the knowledge of language, but can be extended to the whole of cognition. Let us assume that

(10) x = cognition
y = human cognitive behaviour
P = is organized in a modular way

Then, the application of (8) to (9) yields the generalized modularity hypothesis we introduced in Section 2.2.1:

(11) Human cognitive behaviour is organized in a modular way.

(11) is of great importance: although the argumentation which it implicitly rests on and which we have just explicated appears to be very simple, it leads to far-reaching consequences. In what follows let us focus on those from which we might expect an answer to $(P1')_{ch13}$. In order to do so, it is useful, first of all, to remind the reader of the basic categories which are related to (11) and which we explained in (1)–(17) in Section 2.2.1. Our later argumentation will use these as background assumptions.

Now, let us reconsider the argument which we introduced in (18) in Section 2.2.2 and which leads from (11) to relevant metascientific conclusions such as (MMH′) repeated here as (12c). The argument is this:

(12) *If*
a. human cognitive behaviour is modularly organized, and
b. scientific knowledge is part of human cognitive behaviour,
then
c. scientific knowledge is modularly organized.

Two observations present themselves immediately. Firstly, the structure of the inference applied in (12) seems to be related to that of (8), the main difference being that — in opposition to the latter — it is not a property of the whole that is inferred from a property of the part, but vice versa:[11]

(13) *If*
a. *x* is part of *y*, and
b. *y* has the property *P*,
then
c. *x* also has the property *P*.

Secondly, (12c) is assumed to be an empirical hypothesis which a possible *modular approach to the cognitive science of science* can be built on. Such a modular approach to the cognitive science of science can be minimally characterized like this:

(14) A *modular approach to the cognitive science of science*
a. accepts (12c) as its empirical hypothesis;
b. interprets (12c) in terms of (1)–(17) in Section 2.2.1;
c. is not of a philosophical, but of an empirical, scientific nature.

Before applying this modular approach to the cognitive science of science to the problem of the Criterion, as the next step of our argumentation, we have to show how such an approach can capture standard problems of metascience.

13.2.3 An empirical solution to the problem of the Criterion

Let us remember that the problem of the Criterion as outlined in Section 13.2.1 was the following. Why do we accept sentence *a* as true? According to the first possible answer the reason we accept the truth of *a* is that we obtained *a* by the use of a certain method *m*. But why did we apply *m*? If we try to answer this question on the basis of (12c), we may avoid both circular reasoning and the infinite regress. In particular, we have to argue in a way which is fundamentally different from the way traditional philosophy would argue. The research strategy of a modular approach to the cognitive science of science outlined in Section 8.2.2, yields a clear answer. The reason why we applied *m* is that the specific interaction of the universal principles of human cognitive behaviour yields exactly those specific methodological rules which *m* rests on and not others. Should we carry out the mechanism described in 8.2.2 whose workability has already been demonstrated in Section 8.3, within the frames of a modular approach to the cognitive science of science *neither an infinite regress nor circularity would occur*, because our task is to describe those principles and parameters which underlie *m* and, on the basis of such a *description*, we may also achieve a metascientific *explanation* of why *m* is applied.

This seemingly simple solution to the problem which we could obtain in analogy to what has been carried out in Section 8.3 is to be evaluated along the lines of the following considerations:[12]

i. *The modification of the problem by the addition of a new condition.* On the one hand, it goes without saying that the above argumentation yields a possible solution to the problem of the Criterion: if (14) is given, both circularity and the infinite regress can indeed be avoided. On the other hand, however, it is easy to realize that what we have solved is *not* the original problem, but rather, *its empirical reformulation.* The main difference between the original problem raised by Sextus Empiricus and the one we have solved is that the former did not include (14) as a backgound assumption while the latter did. Therefore, we solved a problem which is not independent of that of the Criterion, but which is not identical with it, either. It consists of two parts: firstly, a reformulation of Sextus Empiricus' problem which was presented in Section 13.2.1; and secondly, of (14), which the original formulation of the problem did not contain. The finding that a possible modular approach to the cognitive science of science yields a solution to a problem which is the empirical reformulation of a classic philosophical quandary should not be surprising. On the contrary: this procedure corresponds exactly to what, due to (1), is to be expected from cognitive linguistics in general and cognitive semantics in particular.

ii. *Explanation instead of justification.* The analytic philosophy of science intends to *justify* scientific knowledge about the world. It strives to answer questions like "Why *a*?" and "Why *m*?" by justifying *a* or *m*, respectively. However, if we develop a metatheory on the basis of the empirical hypothesis (12c), we are aware of the fact that this metatheory must be of an empirical nature (see (14c)). Since it is presupposed at the outset that empirical theories do not aim at the justification of their object of investigation, but rather, at its description and explanation, the solution proposed by relying on (12c) differs from the objectives of traditional philosophy considerably. In particular, it interprets why-questions not as questions which are answered by justifying our claims, but as questions which may be answered by the explanation of the facts described. The structure of the corresponding argumentation can be rendered like this:

(15) a. *Classic philosophical problem:* Why do we accept the truth of *a*?
  b. *The reformulation of* (15a) *as an empirical metascientific problem:* Why do we accept the truth of *a*, *if* scientific knowledge is, according to (12c), organized in a modular way?
  c. *Empirical answer to* (15b): We accept *a* as true, because we obtained it by the application of the method *m*. But why did we apply *m*? The reason we applied *m* is that the specific interaction of universal principles belonging to different modules of human behaviour yields the very methodological rules which correspond to *m* and not others.

As a result of this typical cognitive scientific argumentation, it is plausible to accept the empirical reformulation and solution of the problem of the Criterion within the framework of a modular approach to the cognitive science of science.[13] *Thus we obtained the solution to $(P1')_{ch13}$:*

$(SP1')_{ch13}$ If the metascientific extension of the generalized modularity hypothesis (MMH′) were accepted, the classic philosophical problem of the Criterion of truth could be reformulated and solved empirically.

However, if we want to consider $(P1)_{ch13}$, it is not sufficient to focus on these *prospects* of modular cognitive semantics; we also have to shed light on its *limits* as well. It is this which the next section will be devoted to.

13.2.4 The limits of the solution

Without striving for completeness, at least the following two remarks seem to be important.

i. *Plausible inferences.* The reasoning which led from (7) to (9) made use of (8) as a typical inference applied in cognitive science. This fact is relevant, because (8) is a *plausible*, that is a *non-demonstrative* inference. The main difference between demonstrative and plausible inferences is — as we have seen in Section 11.3 — that, while in the former case true premises yield true conclusions, plausible inferences do not guarantee the truth of the conclusion reached. Even if the premises are true and we apply the inference correctly, the premises only contribute to the *credibility* of the conclusion. Thus from the assumption that x is part of y and x has the property P it may, but need not follow that y also has the property P. Therefore, any problem-solving strategy relying on an inference like (8) is uncertain and fallible.

Likewise, (13) is a plausible inference as well. Accordingly, this kind of fallibility is characteristic not only of the way cognitive scientists infer (9) from (7) on the basis of (8), but also of the way we obtained (12c) — that is (MMH′) — from (11) by the help of (13). Since the empirical solution to the problem of the Criterion suggested above was the result of a series of argumentative steps which rest on plausible inferences, the solution itself is burdened with all the uncertainty and fallibility of plausible argumentation.

In Section 11.4 we remarked that in certain cases it is difficult to distinguish between informal fallacies and plausible inferences because the structure of the inference itself is not enough to delimit these two sorts of non-demonstrative reasoning (see the quotation in Section 11.3). It is important to remark that the inferential patterns in (8) and (13) are often regarded as fallacious — but not in every case: "Inference from a property of the part to a property of the whole is warranted only in some cases" (Walton 1997: 213.). It is the *context* of argumentation which makes this pattern appear a plausible inference via the fact that the whole-part relationship is assumed to be relevant. Should we not accept the plausible nature of (8) in the context of the cognitive scientific theories in which it is used, then one of the major arguments of cognitive science should be rejected for logical reasons along with theories such as Chomsky's theory of language and cognition (see also Forrai 1987), or the two-level approach. Nevertheless, this applies of course to (13) as well and suggests that the way the basic hypothesis (MMH′) of a modular cognitive semantic approach to the cognitive science of science was inferred in Section 2.2.2 is really *on the boundary* between plausible reasoning and fallacious inferences. *This indicates a substantial weakness of such an approach.*

ii. *An antinomy.* As is well known, cognitive semantics is far from being unified; on the contrary, it consists of several, essentially different, often antagonistic theories presupposing incompatible empirical hypotheses. At the outset none of these can be regarded as an effective and successful tool for solving the philosophical problems raised. For example, we could have attempted to choose the holistic hypothesis instead of the modularity hypothesis as a starting point in order to tackle the problem of the Criterion.

If we accept this possible pluralism of both cognitive semantics and, as a consequence, of metatheories construed from cognitive semantic hypotheses, we are forced to confront serious questions. On the basis of what criteria, what aspects, what kind of theoretical or empirical argumentation can we select precisely that framework which will serve as the starting point for the construction of a metatheory capable of tackling classic philosophical problems, such as the problem of the Criterion? If, as we have just seen, we obtain such a metatheory by way of plausible argumentation, how can we decide between alternative hypotheses?

No doubt, these questions are not to be answered on the objectscientific level, i.e. on the level of cognitive semantic hypotheses and of the cognitive semantic theories maintaining them. Nevertheless, they cannot be answered one level higher, i.e. on the metascientific level, either. This is because in order to obtain the answers we have to compare the effectiveness, the workability and the explanatory potential of the metatheories at our disposal. However, the precondition for such a comparison is their description. The description of metatheories means that we have to locate our descriptive apparatus yet one level higher, namely, on the *meta-metascientific* level: and so on *regressus ad infinitum.* The unavoidable conclusion is, therefore, that the task of accounting for the classic philosophical problem of the Criterion by applying the empirical methods of a cognitive semantic theory yields a similar problem which *involves a fallacy.*[14]

13.2.5 Conclusions

The case study developed was motivated by the question $(P1')_{ch13}$, which asked whether the two-level approach was capable of an empirical solution to the classic philosophical problem of the Criterion. From the example of modular cognitive semantics we have seen that the answer is definitively *affirmative.* However, the case study has also shown that, although cognitive semantics has a potential going beyond that of other branches of linguistics, its *limits* are clearly set:

$(SP1)_{ch13}$ As witnessed by a case study, the two-level approach was capable of solving a classic philosophical problem. At the same time, however, it cannot transgress certain limits. In particular, modular cognitive semantic argumentation is burdened, among other things, with the peculiarities of plausible inferences and the emergence of fallacies.

At this point it is important to remind the reader of the fact that the use of plausible inferences and the emergence of fallacies have to be evaluated differently and they are related to very different kinds of limits. Plausible inferences are considerably *weaker* than deductive ones: they undoubtedly indicate a limit which they cannot transgress, namely that between the credibility and the certainty of conclusions. Nevertheless, the use of plausible inferences in scientific reasoning definitely *does not* count as a mistake; on the contrary, as emphasized in Section 11.3, plausible inferences are indispensable tools of both everyday thinking and scientific inquiry. As opposed to the possible fruitfulness of plausible inferences, fallacies are serious *shortcomings*: they indicate the limit between correct and incorrect reasoning.

13.3 Case study 2: Holism

13.3.1 Zeno's Arrow

By analogy to our argumentation in Section 13.2.1, so as to test holism, now we have to find another classic philosophical problem to which $(P2)_{ch13}$ may be reduced. This is a relatively easy task, because, unlike the proponents of the two-level approach, Lakoff and Johnson discuss numerous examples in their *Philosophy in the Flesh.*[15] One example which is of the same fundamental nature as the problem of the Criterion and is analyzed by the authors in an explicit way is Zeno's Arrow. Lakoff and Johnson raise the problem of Zeno's Arrow as follows:

> Suppose, Zeno argues, that time really is a sequence constituting a time line. Consider the flight of an arrow. At any point in time, the arrow is at some fixed location. At a later point, it is at another fixed location. The flight of the arrow would be like the sequence of still frames that make up a move. Since the arrow is located at a single fixed place every time, where, asks Zeno, is the motion? Time, Zeno argues, is not divided up into instants. (Lakoff and Johnson 1999: 157.)

Thus, in the light of this formulation Zeno's classic standpoint is this:

(16) a. *Classic philosophical problem:* Is there motion?
b. *Classic philosophical solution:* There is no motion, because at any point of time the arrow is at some fixed location.

Therefore, in the case study to follow we reduce $(P2)_{ch13}$ to $(P2')_{ch13}$:

$(P2')_{ch13}$ Could the classic philosophical problem of the Criterion be empirically reformulated and solved, if the metascientific extension of the main hypothesis of the cognitive theory of metaphor (MHH') were accepted?

13.3.2 An empirical solution to the problem of Zeno's Arrow

Lakoff and Johnson specify the task of their own version of cognitive semantics as follows:

> The meaning of any philosophical question depends on what conceptual system is being used to comprehend the question. That is an empirical issue, an issue to be taken up by cognitive science in general and cognitive semantics in particular. […] The same is the case for any proposed answer. An answer to a question like 'What is time?' is given relative to a philosophical conceptual system in which that

> answer is a meaningful answer. *Such a philosophical conceptual system is part of the conceptual system of the philosophers doing the inquiry.* The conceptual systems of philosophers are no more consciously accessible than those of anyone else. To understand what counts as a meaningful answer, *one must study the conceptual systems of the philosophers engaged in that inquiry.* That too is an empirical question for cognitive science and *cognitive semantics.* (Lakoff and Johnson 1999: 136; emphasis by underlining in the original, italics added.)

The implicit argumentation underlying this quotation can be reconstructed in the following way:

(17) *If*
- a. "a philosophical conceptual system is part of the conceptual system of the philosophers doing the inquiry",
- b. the conceptual system of the philosophers doing the inquiry belongs to the subject matter of cognitive semantics,

then
- c. a philosophical conceptual system belongs to the subject matter of cognitive semantics.

This argument rests on *a pattern of plausible inference* which will be discussed in Section 13.3.3.

Lakoff and Johnson's solution to Zeno's problem uses a second inference as well:

(18) *If*
- a. "a philosophical conceptual system is part of the conceptual system of the philosophers doing the inquiry",
- b. the conceptual system of the philosophers doing the inquiry is a human conceptual system,
- c. "the human conceptual system is metaphorically structured" (Lakoff and Johnson 1980a: 6),

then
- d. a philosophical conceptual system is metaphorically structured.

So, from (17c) and (18d) it follows that it is the metaphorical structure of Zeno's conceptual system which is responsible for the problem of the Arrow. Therefore, this metaphorical structure needs to be revealed. The conceptual metaphor which is assumed to play a constitutive role in the structure of such a conceptual system is one that may be called the *Moving Observer metaphor* (Lakoff and Johnson 1999: 146):

(19) LOCATIONS ON OBSERVER'S PATH OF MOTION ARE TIMES
THE MOTION OF THE OBSERVER IS THE 'PASSAGE' OF TIME
THE DISTANCE MOVED BY THE OBSERVER IS THE AMOUNT OF TIME PASSED

This metaphor implies that "what we will encounter in the future is what we are moving towards", "what we are encountering now is what we are moving by", and "what we encountered in the past is what we moved past" (Lakoff and Johnson 1999: 152). It is the identification of the Moving Observer metaphor that yields the empirical solution to the problem of the Arrow which we quote in full:

> [...] In our terms, the idea that time is a linear sequence of points is metaphorical, a consequence of times seen as locations in the Moving Observer metaphor. The mistake, once again, is to take what is metaphorical as literal.
> Incidentally, a cognitive response to Zeno's paradox of the arrow is simple. There is a part of the brain that detects motion. Our motion detectors identify the arrow as moving. That is, our brains give us multiple ways of perceiving and conceptualizing the world. Motion is not a metaphorical concept. The idea is that time is a linear sequence of finite points. Our direct nonmetaphorically structured experience provides a simple response: Of course the arrow is moving. But in addition, we have an unconscious metaphorical conceptualization of instants of time as locations in space. We use this, for example, when we comprehend a picture of a moving object at a time: 'This is Sam driving by directly in front of our house at 10:06 p.m.' In other words, we have more than one way to conceptualize motion — one literal and one metaphorical. We can conceptualize motion directly, as when we think of Sam driving by and the hands of the clock moving. We can also conceptualize motion using a metaphorical conceptualization of time as a line with point locations on it. In the metaphor, and only in the metaphor, there is temporal location. Relative to the metaphor, we can fix a point location in time. Within the metaphor, *at* that point location, there can be no motion, since motion can only occur *over* regions of time in the metaphor. The appearance of paradox comes from attributing real existence to metaphorical point locations. Zeno's brilliance was to concoct an example that forced a contradiction upon us: literal motion and motion metaphorically conceptualized as a sequence of fixed locations at fixed points in time. (Lakoff and Johnson 1999: 157–158; emphasis in the original)

On the basis of this quotation, the way Lakoff and Johnson proceed is analogous both to Chomsky's handling of the Meno-paradox and our treatment of the problem of the Criterion in the previous section.That is:

(20) a. *Empirical reformulation of the classic philosophical problem:*
Is there motion, *if*
– there is a part in the brain that detects motion, and

– our motion detectors identify the arrow as moving?

b. *Empirical solution to the empirical problem:*
There is motion, *because*
– our brains give us multiple ways of perceiving and conceptualizing the world and, within the latter, motion as well,
– we have a literal and a metaphorical way of conceptualizing motion,
– (16b) is false, because Zeno committed the mistake of taking the Moving Observer metaphor to be literal.

From (20) we obtain the following answer to $(P2')_{ch13}$.

$(SP2')_{ch13}$ If the metascientific extension of the main hypothesis of the cognitive theory of metaphor (MHH′) were accepted, the classic philosophical problem of Zeno's Arrow could be reformulated and solved empirically.

13.3.3 The limits of the solution

The aim of this section is not to review the criticism which the cognitive theory of metaphor has been confronted with during the past two decades and which is well-known from the literature.[16] Rather, it will be sufficient to mention just three aspects which seem to undermine $(SP2')_{ch13}$ in a relevant way.

i. *Plausible inferences.* What we have said about the use of plausible inferences with respect to the metascientific extension of modular cognitive semantics also applies to the reasoning that led to Lakoff and Johnson's solution to Zeno's antinomy. (17) and (18), which played a central role in the argumentation, apply to the pattern of inference in (13). The latter is, at best, a plausible inference which, in a different context may even count as a fallacy. Therefore, the argumentation on which the solution rests is fallible.

ii. *The holistic aporia.* Müller (1991) reveals a problem which he calls the 'holistic aporia' and which may shed light on some hidden but quite interesting aspects of the inferences put forward in (17) and (18). Müller writes:

> If the holistic demand is really taken seriously and followed to its logical conclusion, it results in an aporia. For example, a complete and holistic theory of semantics which is an indispensable component of an accomplished theory of language, tends to comprise along with the 'knowledge of the world' the whole world as well and would in its last consequence yield a miniature model of the cosmos.(Müller 1991: 406 f.; my translation, A. K.)

It seems to be the case that the holistic aporia is a serious problem and therefore, although implicitly, the cognitive theory of metaphor tries to avoid it. The only way to avoid it is, of course, the subdivision of the subject matter of the theory into 'parts' and indeed this is precisely what happens: in particular, (17) rests on an inference the first premise of which states a whole-part relationship exemplified by the *first quotation* in Section 13.3.2. This pattern of inference suggests that, so as to avoid the holistic aporia as a potential consequence of the holistic stance, there is no avoiding the subdivision of the subject matter of the theory into what Lakoff and Johnson explicitly call "parts of". However, the principles according to which this is carried out are anything but clear. Thus the inference in (17) is — apart from what has been said in (i) in this section, also because of the unclarified and highly uncertain status of premise (a) within the theory — only plausible, at best. Nevertheless, it cannot be excluded at the outset that further inspection would reveal that this premise is the result of the internal inconsistency of the approach. Obviously, as long as the relationship between this kind of subdivision into parts and the assumed 'wholeness' of the subject matter remains unclear, the fallacy of inconsistency cannot be excluded at the outset.

iii. *An antinomy.* First of all, we obtain the same sort of antinomy which has already been mentioned in (ii) in Section 13.2.4 involving fallacious reasoning. Lakoff and Johnson's cognitive theory of metaphor is, of course, not the only approach to cognitive semantics. Clearly, the notion of cognitive semantics is applied to several very different approaches which may even be incompatible with each other. Therefore — as an immediate metatheoretical consequence of Quine's underdetermination thesis — on the basis of the same 'empirical data' we may in certain cases construe metatheories contradicting each other.[17]

However, this antinomy may have further consequences. For example, if we accept the cognitive theory of metaphor, then we have to assume that metaphor is a central tool of conceptualizing the world. This implies, furthermore, that we may consciously and deliberately use this potential of metaphors to develop scientific theories,[18] or to solve philosophical problems. For example, in Section 9.2.2 we quoted Liebert (1995: 439 and 445) on a possible application of the cognitive theory of metaphor. Liebert argued that revealing the metaphorical structure of scientific discourse could stimulate research in that researchers could make conscious use of metaphors. Consequently, the *meta*scientific application of the cognitive theory of metaphor may contribute to the solving of the *object*scientific problems which researchers are after. This is what we called 'constructivity' in Sections 9.3 and 11.2.

No doubt, this view implies a clear difference between the metascientific application of the cognitive theory of metaphor and the way the analytic philosophy proceeds, because the latter maintains the autonomy of both metascientific reflexion and philosophical analysis with respect to objectscientific research, whereas the former holds that objectscientific research may be influenced constructively by the results of metascientific reflexion and philosophy.

But at this point a serious problem arises. Namely, if we accepted another approach to cognitive semantics as a possible source of the cognitive science of science which is not compatible with the cognitive theory of metaphor, then we would have to draw the conclusion that we should construct scientific theories or philosophical systems whose terms are basicially not metaphorical. So, depending on which approach to cognitive semantics we want to use for the metascientific analysis of theory formation or the solution of philosophical problems, we will obtain very different accounts of the nature of theory formation or possible solutions to philosophical problems. But, depending on the nature of this picture, our account will suggest incompatible theoretical aims and methodological principles to the working scientist or philosopher.

13.3.4 Conclusions

By analogy to Section 13.2, the case study developed in Section 13.3 was motivated by the question $(P2')_{ch13}$. The application of the cognitive theory of metaphor as outined in Lakoff and Johnson (1999) showed that the answer is clearly affirmative. Nevertheless, the limits of this approach became quite evident as well. Thus we obtain the following solution to $(SP2)_{ch13}$.

$(SP2)_{ch13}$ As witnessed by a case study, the cognitive theory of metaphor was capable of solving a classic philosophical problem. At the same time, however, it cannot transgress certain limits. In particular, holistic cognitive semantic argumentation is burdened, among other things, with the peculiarities of plausible inferences and the emergence of fallacies.

13.4 Summary

In the present chapter we started from the assumption that one central aim of cognitive science in general and cognitive semantics in particular was the empirical reformulation and empirical solution of classic philosophical problems. This thesis was tested with the help of two case studies. Firstly, the workability of the two-level approach was analyzed with respect to the problem of the Criterion. Secondly, Lakoff and Johnson's solution to the problem of the Arrow was evaluated. The results of both case studies seemed to be generalizable: although cognitive semantics may tackle classic philosophical problems and although it may offer novel empirical solutions to them, the results are not significantly better than those of traditional philosophical approaches for they yield antinomies involving fallacies and are exposed to the weaknesses of plausible inferences.

Chapter 14

Conclusions: Limits

14.1 Introduction

In Chapter 11 we asked three questions with respect to the possible limits of the metascientific application of the two-level approach and the cognitive theory of metaphor. The case studies in Chapters 12 and 13 gave us clear-cut answers. In analogy to the way we proceeded in Chapter 10, the present chapter asks the following question:

$(P)_{ch14}$ By what further limits can the findings of Chapters 12 and 13 be supplemented?

We will try to answer this question in two steps. It is worth emphasizing again that the case studies which play a central role in our attempt to solve the main problem of the book were *deliberately* constructed in such a way that they exemplify not only the possibility of new applications of cognitive semantics, but also some of their shortcomings. Accordingly, in Sections 14.2 and 14.3 we will mention a few instructive examples of obvious shortcomings *of the two-level approach or the cognitive theory of metaphor*. Thereby emphasis is laid on the way they are *applied* in the course of the case studies. We will neither strive to give an overview of the criticism from the literature, nor do we enumerate all the unsolved problems which our analyses yield. Rather, we will restrict our attention to certain *examples* which seem to be *instructive* from the point of view of our main problem (Q‴).

In Section 14.4 we will point out that it is not only the internal shortcomings of the two approaches at issue that set limits to their applications, but also *certain general properties of the case studies*. Section 14.5 will present the solution to $(P)_{ch14}$.

14.2 On certain limits of the cognitive theory of metaphor

To show that the conclusions we have drawn from the case study in *Chapter 4* may be seriously undermined, we will touch on one aspect of the problem of metaphor identification: the possibility of *alternative analyses.* The source of the problem is that, in accordance with (20b) in Section 2.3.1 it is metaphorical expressions as linguistic structures which are claimed to serve as the evidence for identifying conceptual metaphors. However, as for example McGlone points out,

> a [...] problem with the linguistic evidence is that although it may be consistent with (if not force) the metaphoric representation claim, it is nonetheless misleading. Our intuitions about how idioms metaphorically acquire their meanings are often quite compelling, even when they are *dead wrong.* The very act of generating an intuition about an idiom's meaning can make one resistant to alternative accounts that may in fact be correct. (McGlone 2001: 95 ff.; emphasis added)

With respect to this problem of alternative analyses the theoretical terms used in the Standard Theory and Government-Binding Theory behave analogously to the idioms mentioned in the quotation. Let us remember that in Chapter 4 theoretical terms were treated as (parts of) metaphorical expressions — i.e. as linguistic structures — and on the basis of these we proposed the conceptual metaphors mentioned in (2)–(5) and (7)–(8).

In identifying the conceptual metaphors that within the framework of the cognitive theory of metaphor are assumed to underlie the theoretical terms of generative grammar, we started from considerations put forward in Riley (1987). Nevertheless, in accordance with the above quotation our analyses suggested in the case study may be "dead wrong" and alternative analyses could also have been proposed. For example, the following ones seem to be possible as well as alternatives to (2)–(5) and (7)–(8) in Chapter 4.[1]

In the Standard Theory the basic metaphors are LINGUISTIC EXPRESSIONS ARE OBJECTS (EITHER LIVING ORGANISMS OR INANIMATE OBJECTS) and A SENTENCE IS A COMPLEX OF OBJECTS (EITHER A PLANT-TREE OR A FAMILY). The basic assumption then is that the expressions that make up a sentence are in a hierarchical relationship. So, there must be a conceptual metaphor ABSTRACT LINGUISTIC HIERARCHICAL STRUCTURE IS PHYSICAL OR SOCIAL STRUCTURE.

Against this background, the following specific conceptual metaphors seem to present themselves:

(2′) THE HIERARCHICAL STRUCTURE OF A SENTENCE IS THE PHYSICAL 'PART-OF' STRUCTURE OF A TREE (TREES CONSISTING OF BRANCHES, etc.)

Given (2′), it follows that

(3′) CHANGES IN THE HIERARCHICAL STRUCTURE OF A SENTENCE ARE PHYSICAL MANIPULATIONS OF OBJECTS

Moreover:

(4′) THE HIERARCHICAL STRUCTURE OF A SENTENCE IS THE SOCIAL ORDERING OF A FAMILY (PARENTS HAVING CONTROL OVER CHILDREN)

(5′) LEVELS OF SENTENCE STRUCTURE ARE PHYSICAL LAYERS OF A CURRENT (OF LIQUID)

In turn, an alternative analysis of the conceptual metaphors in Government-Binding Theory may be based on the basic metaphor ABSTRACT FORMAL / STRUCTURAL CONTROL IS SOCIAL / PHYSICAL CONTROL.

In this case, the specific metaphors are:

(7′) ABSTRACT RELATIONS OF FORMAL / STRUCTURAL CONTROL AMONG ELEMENTS OF A SENTENCE ARE SOCIAL RELATIONS AMONG PEOPLE (IN A SOCIETY)

(8′) ABSTRACT RELATIONS OF FORMAL / STRUCTURAL CONTROL AMONG ELEMENTS OF A SENTENCE ARE CASES OF PHYSICAL CONTROL AMONG PHYSICAL OBJECTS

Now, the point is that if we accept (2′)–(5′) and (7′)–(8′) instead of (2)–(5) and (7)–(8) respectively, then it follows that the main difference between the Standard Theory and Government-Binding Theory is this: while the former focuses on the hierarchical structure of the components of sentences, Government-Binding Theory emphasizes that certain elements control other elements. This result is not identical with the conclusion we drew in Section 4.2:

> Theoretical terms in the Standard Theory are rooted in metaphorical concepts which conceptualize grammar with respect to *creativity, dynamism, motion, growth* (cf. also Riley 1987). Theoretical terms in Government-Binding Theory are based on metaphorical concepts which conceptualize grammar as being something which is exposed to *limitations, restrictions, confinements* (cf. also Riley 1987, Junker 1992).

In sum, the attempt to answer $(P)_{ch4}$ by arguing for $(SP)_{ch4}$ revealed the possibility of alternative analyses which, accordingly, yield alternative conclusions concerning the nature of theoretical terms while resting on the same metaphorical expressions. No doubt, this is a serious limit of the approach.

After these relatively detailed considerations exemplifying basic difficulties which the metascientific application of the cognitive theory of metaphor brought to the surface, it will be sufficient to point out briefly another serious difficulty with respect to *Chapter 9*. In particular, the way conversation analysis and the cognitive theory of metaphor were integrated in the case study as proposed by Liebert may be interpreted as circular. On the one hand, the case study shows that scientists are capable of making their metaphors fit their scientific ideas; on the other hand, however, it is assumed that these metaphors are responsible for the nature of the ideas themselves. In fact, this is a serious limit of $(SP1)_{ch9}$ and $(SP2)_{ch9}$, which ought to be avoided by resolving the circularity.

14.3 On certain limits of the two-level approach

The second quotation from Bierwisch (1983: 67) in Section 2.1 really seems to suggest that the two-level approach intends to treat theoretical terms in linguistics in the way exemplified in *Chapter 5.* Nevertheless, the question is, whether the relations that hold between different interpretations of the term *command* are indeed of the same sort which the two-level approach focuses on when speaking of 'conceptual shift' or 'conceptual specification'. In particular, several of the interpretations of *command* differ from each other insofar as one of them contains an additional condition which another does not contain. But this hierarchical relationship seems not to be identical with the kind of relations exemplified in (13) and (14) in Section 2.2.1. One of the main reasons for this difficulty is that the two-level approach does not clarify what is meant by 'conceptual domain'. Thus it is not clear what relations hold between the 'conceptual domains' connected by conceptual shift and how to identify the 'conceptual domains' within which conceptual specification works.

Another problem is how to differentiate between the 'vagueness' and the 'semantic underdetermination' of theoretical terms in linguistics. For example, in the case of the term *sentence* mentioned in Section 5.2.4 or the term *verb* there are many phenomena which suggest that it is not clear whether certain structures of natural language belong to the domain of such a term or not; but these cases may also be analyzed as being based on 'conceptual shift'. For example, it may be an open question whether a certain lexical unit is a verb or an adjective. In this case the term *verb* seems to be rooted in vagueness rather than in semantic underdetermination. Analogously, it may be an open question whether a certain syntactic unit is a *sentence* or a *phrase* or even a *text.* The fact that often it is not self-evident whether we are dealing with 'vagueness' or with 'semantic underdetermination' with respect to theoretical terms, suggests that the pre-explicative nature of the basic notions of the two-level approach itself seriously affects the evaluation of the applications. Namely, if, due to the uncertainity of expressions such as 'conceptual shift' it may be the case that phenomena which we have analyzed as instances of 'conceptual shift' turn out to be rooted in 'vagueness', then the conclusions drawn in Chapter 5 may be highly questionable. The same applies to nearly all central terms of the two-level approach such as 'concept', 'module', 'principle', 'representation' etc.

In sum, the attempt to answer $(P)_{ch5}$ by arguing for $(SP)_{ch5}$ revealed the weaknesses of the basic notions of the two-level approach as one of its serious limits. Although here we do not strive to recapitulate the critical literature on the two-level approach, in this context it is instructive to add the following remarks.

With respect to the cognitive theory of metaphor the terminological uncertainty is understandable, because it consciously rejects the analytical philosophy of science which focuses, among other things, on the necessity of precise definitions in theoretical contexts. In contrast, the two-level approach conforms to the tradition of the analytic philosophy of science, and presupposes applications within artificial intelligence research (see e.g. Lang and Carstensen 1990, Lang et al. 1991). Therefore, the pre-explicative nature of its basic terms such as 'module', 'representation', 'concept', 'literal', 'conceptual shift', 'principle' etc. and the quasi-formal analyses we exemplified in Chapter 2 seem to be more surprising. The following general criticism, which Carstensen interprets to apply to the two-level approach as well, is very instructive:

> In reality, the work of a computational semanticist as a 'common-sense semanticist' will basically consist in performing a semantic analysis of certain linguistic constructions which is plausible or intuitively evident for him and either done by introspection or based on common psychological opinion, that is, exclusively with the aid of certain general or possibly contextspecific [...] technical terms — that is, based *on extremely vague notions*, and that means here: notions which have not been subject to a formal reconstruction or logical explication so far [...] (Werner 1985: 494; quoted in Carstensen 1995: 119–120; emphasis added).

In commenting on this quotation Carstensen writes:

> Alas, I have to reply, a vague but *inadequate* notion will not become better through formalization unless a heuristic loop or interaction of theory building and formalization is tacitly assumed. (Carstensen 1995: 120; emphasis in the original)

The open question for us then is whether the limits we have just pointed out make the two-level approach appear to be inadequate for its application to the structure of scientific terms or not.

Let us turn to *Chapter 8*. As in the previous cases, here we also do not strive to give a comprehensive list of counter-arguments against the considerations which led to $(SP1)_{ch8}$ and $(SP2)_{ch8}$; rather, we focus on some crucial remarks which, by exemplifying the metascientific application of the two-level approach, might be instructive in revealing some of its limits.

Firstly, in the case study very much depends on the assumption that there is a set of modules, and among these, a social one which consists of two submodules, namely, the interactional and the motivational submodule. Without assuming that there is an autonomus system of motivational representations, motivational rules and motivational principles, none of the later steps of our argumentation can be accepted. However, proving the existence of relatively autonomous systems and subsystems is one of the most difficult tasks of modularism, because *there is a substantial lack of empirical evidence.* This is instructively documented for example in Bierwisch and Lang (1989):

> [...] it is clear that *our statements on the autonomy and interaction of the components* are, for more than one reason, of a predominantly *heuristic* nature. [...] The justification of our more general statements is therefore based chiefly on the fact that they provide the framework for the systematic explanation of a relatively complex, structured part of the relevant facts and make it possible to examine them in the light of the theory internal and external relations. (Bierwisch and Lang 1989: 495–496; emphasis added)

In the work cited these remarks concern mainly the status of semantic representation. It is instructive to remark that the conclusions just quoted were drawn in spite of the fact that over several hundred pages an attempt was made to find arguments for the relative autonomy of semantic representations. Thus, the implications of the quotation may be even more serious if they are related to the motivational module, because there is no evidence at all proving that it is a relatively autonomous system. Consequently, *all* the arguments which seem to speak for the acceptability of the case study carried out in Chapter 8 rest on heuristic considerations. Therefore, by the example of its metascientific application the case study demonstrates very clearly that modular cognitive semantics cannot trangress the limit drawn by the strongly hypothetical and heuristic nature of the very modules which are indispensable starting points for any kind of argumentation within modular cognitive semantics.

The case study also serves to exemplify a *second* but related aspect of heuristic argumentation which is very instructive. In particular, in one of his classic works Polya characterizes heuristic argumentation, among other things, as follows:

> Heuristic reasoning is reasoning not regarded as final and strict but as provisional and plausible only, whose purpose is to discover the solution of the present problem. (Polya 1954: 102)

> In the *heuristic* syllogism the premises constitute only one part of the basis on which the conclusion rests, the fully expressed, the 'visible' part of the basis; there is an unexpressed, *invisible* part [...]. In fact, it can happen that we receive some new information that leaves our belief in both premises completely intact, but influences the trust we put in *A* in a way just opposite to that expressed in the conclusion. To find *A* more plausible on the ground of the premises of our heuristic syllogism is only reasonable. Yet tomorrow I may find grounds, not interferring at all with these premises, that make *A* appear less plausible, or even definitively refute it. *The conclusion may be shaken or even overturned completely by commotions in the invisible parts of its foundation, although the premises, the visible part, stand quite firm.* (Polya 1954: 223–224; emphasis by italics added)

The phenomena which this quotation touches on may be rich and diverse, depending on the particular field of research and the particular context of argumentation. The case study in Chapter 8 is a specific example of how the 'invisible' basis may work and what its consequences are. In the case study we attempted a sociological extension of the two-level approach by integrating it with an approach to the sociology of knowledge. In the course of the integration of these two approaches we tried to put forward coherent and consistent solutions to $(P1)_{ch8}$ and $(P2)_{ch8}$ by arguing heuristically. Nevertheless, the background information lying behind each of the two theories is very complex and, in several respects, also very different. So, although at every point of the argumentation we tried to identify the major premises and to explicate the way we argued on their basis, we must not exclude the possibility that *there is indeed an invisible part of the basis.*[2] This invisible part involves those assumptions of the two approaches which we did not make use of in the argumentation and their indirect consequences which we have not yet realized. So, even if at first sight the results of our argumentation may seem plausible, our conclusions "may be shaken or even overturned completely by commotions in the invisible parts of its foundation".

By way of summary, we may conclude that the relevance of $(SP1)_{ch8}$ and $(SP2)_{ch8}$ is limited, among other things, because (a) the modules whose existence the argumentation hinges on are heuristically motivated theoretical constructs, and (b) apart from the 'visible basis' of the argumentation there is an 'invisible basis' whose implications may be different from those of the visible one.

14.4 The limits of the case studies as thought experiments

The examples discussed in the previous two sections touch on limits which are *inherent* in the two approaches, but which have been brought to surface by *the particular applications* we carried out. However, there are certain limits which are set by *certain general properties of the case studies.*

Since both the cognitive theory of metaphor and the two-level approach consider themselves to be 'empirical', the best way of testing their workability would be, of course, by carrying out 'empirical experiments'. However, neither the generalized modularity hypothesis nor the main hypothesis of the cognitive theory of metaphor have been supported convincingly by 'empirical evidence' (however we choose to interpret 'empirical evidence'). Moreover, the object of our investigation, namely, certain aspects of scientific knowledge is highly complex, whereas the pluralistic state of the situation in the philosophy of science is not capable of providing us with firm points of departure. For these two reasons the cognitive semantician who wants to investigate aspects of scientific knowledge by the metascientific extension of cognitive semantics has *no other choice* than to treat the case studies carried out as *thought experiments.*[3]

Accordingly, as was *consequentially* emphasized, our case studies basically give answers to two kinds of questions:

- *What would be the case if* scientific concept formation in particular and certain aspects of scientific knowledge in general were based on the interaction of modules?
- *What would be the case if* scientific concept formation in particular and certain aspects of scientific knowledge in general were structured by metaphorical processes?

The case studies focusing on such questions reveal the following general limits of the applications:

First of all, it may be the case that the two models which we constructed — i.e. a modular and a holistic model — *ignore* certain relevant features of scientific concept formation in particular and scientific knowledge in general whose inclusion would, however, change the premisses and, in concequence, possibly lead to *conclusions other than* those which have been drawn. The risk of this kind of failure is relatively *high* because our knowledge of how scientific cognition is structured and proceeds, is — owing to the shortcomings of the analytic philosophy of science and the immaturity of the cognitive science of science — modest, deficient and hypothetical.

Secondly, due to the "What if" questions the case studies are constructed by *manipulating* the models — i.e. the modular and the holistic models of scientific cognition, respectively. These manipulations rest mainly, as was shown, on accepting *additional background assumptions.* These additional background assumptions are of a hypothetical nature, because — apart from more substantial reasons — they could not be discussed extensively and could not be proved convincingly; thus, we cannot exclude at the outset that in the light of new information and later considerations they may turn out to be false or at least questionable.

Thirdly, the nature of the *terminology* which the two models made use of may be risky as well. Often it is the case that pre-explicative terms are more fruitful than precise explications. However, in other cases, it may be important to provide precise explications of the terms one uses. Due to well-known difficulties resulting from the state of the art both in cognitive linguistics and the philosophy of science, the basic strategy of the case studies was to use the terms pre-explicatively, except when, for heuristic reasons, their clarification seemed to be both possible and well-motivated. The question, then, is whether our findings would have been considerably different if we had made different decisons as to which terms should be explicated and which used pre-explicatively.

Fourthly, the *consistency* of the case studies resting on the two models thus characterized cannot be maintained at the outset for at least two reasons. The first is this:

> Thought experiments can also fail because a thought experimenter makes a mistake as to whether he has constructed an internally consistent model. Inconsistency may be difficult to spot. (Cooper 1999: 268)

This kind of failure has to be distinguished clearly from another case: in particular, it may be *one of the explicitly formulated tasks* of thought experiments to reveal the possible inconsistency of a given model (see Peijnenburg and Atkinson 2003: 306f.). As regards the first situation, let us remark ironically that it will be the task of the reviewers to check whether the author of this book made mistakes which are responsible for the possible inconsistency of the case studies — certainly it would be beside the point to go into a detailed self-evaluation of this kind here. In opposition to this, with respect to our main problem (Q‴), the second case is of special interest in the present context. As we know, very often the inconsistency of the models can be revealed only by manipulating them *in the course of applications.* In this sense the way we deliberately manipulated the two approaches during their metascientific applications was *successful*: the unsolved difficulties we pointed out in Sections 14.2 and 14.3 *seriously touch on the potential inconsistency of the two approaches and this clearly refers to at least some of those limits we wanted to detect.* Nevertheless, the real or apparent inconsistent nature of scientific approaches does not mean that they have to be discarded automatically. The role of inconsistencies in science is currently the centre of interest, although in this respect the state of research is certainly immature:[4]

> Actual scientific practice and the theories produced within its framework seem to contain crude violations of even the most fundamental methodological rules, among them the [least] controversial one: the law of non-contradiction. [...] All the while, however, internal (and external) consistency has properly been regarded as the most fundamental and the least controversial of all the criteria of ideal science. The question is now: what should be done about this chronic discrepancy between actual and ideal science? (Fehér 1990: 231–232)

There are many ways of handling the inconsistency of scientific theories, and it would be one of the most important tasks for the future to explore these possibilities with respect to the results of the case studies. In fact, this is one of the points at which the line of argumentation presented in this book has, for practial reasons, to be finished, but from which future considerations continuing the present enterprise should start.[5]

Fifthly, on the basis of the case studies it was *not* possible to *take sides* between modular and holistic accounts of scientific cognition. Rather, we tried to point out that these two approaches may serve different heuristic purposes and may lead to different accounts of the phenomena they try to capture.[6] Although the basic hypotheses could be treated analogously and the case studies were constructed parallelly, the *applications* of the two approaches could not be compared systematically. We emphasized that although they can be applied to analogous problems, the way they work and the findings to which they lead are very different. One consequence of this is that the quandary of incommensurability does not arise, but in this way the intertheoretical relations between the approaches could not be clarified.

Finally, since the cognitive science of science intends to be an 'empirical' discipline (whatever 'empirical' means), it is exposed to all the weaknesses and uncertainities of empirical approaches at the outset. However, with respect to the cognitive semantic approaches we discussed, these weaknesses and uncertainties are even more dangerous than in the paradigmatic natural sciences. Throughout the book we emphasized that the term 'empirical' was used in a pre-explicative sense, because it is anything but clear, in what sense the two-level approach and the cognitive theory of metaphor consider themselves to be 'empirical'. Moreover, as witnessed by the overview in Chapter 2 and the case studies carried out, both belong to the group of cognitive semantic approaches which Geeraerts characterizes in this way (although Geeraerts had only holism in mind):

> The methodological situation in present-day Cognitive Linguistics is characterized by the existence of two methodological extremes. On the one hand, there is the idealistic approach [...]. On the other hand, there exist various tendencies to objectivize the methods used in Cognitive Linguistics. [...] Most actual work in Cognitive Linguistics is situated somewhere in-between both extremes, or rather, the practical methodology used is more often *introspective* rather than data-driven, but without the outspoken idealistic commitments [...] (Geeraerts 1999: 164; emphasis added)

Accordingly, as already noted in the *Introduction*, we had no other choice than to treat *case studies* as functional analogues to empirical experiments. Therefore, the "What if" situation we outlined — which is closely connected with the immaturity of the theories at issue and with the shortcomings of our knowledge of the highly complex mechanisms of scientific inquiry — leads to hypotheses which are *even weaker* than empirical investigations practiced by the paradigmatic empirical disciplines. Cooper comes to an analogous conclusion with respect to the philosophy of mind:

> [...] thought experiments in the philosophy of mind are more prone to failure than those in, say, physics because less is known about the features of the actual world being investigated. Quite simply more is known about the behaviour of light or of masses than is known about consciousness or personal identity. (Cooper 1999: 270)

Nevertheless, such an assessment of findings does not render them pointless. On the contrary, it motivates the further refinement of our models so as to minimalize the risks mentioned as far as possible. As Cooper put it:

> [...] there is no reason to think of philosophy based on thought experiments and philosophy based on real experiments as opposed to each other. Rather, the two approaches are complementary. Empirical knowledge can assist in the construction of successful thought experiments. Equally, *well constructed thought experiments can be used to supplement the findings of real experiments by allowing predictions to be made about the outcomes of experiments which have not yet been, and maybe never could be, performed.* (Cooper 1999: 270–271; emphasis added)

In conclusion, the elaboration of cognitive semantic models of certain aspects of scientific knowledge seems to be a *fruitful* enterprise which opens *new perspectives* for understanding how cognition is structured and works. But at the same time such an enterprise is undoubtedly *risky*.

14.5 Summary

The considerations of Sections 14.2–14.4 supplement the findings of Chapters 12 and 13 in a way which boils down to the following summary of the limits of the metascientific application of the two level-approach and the cognitive theory of metaphor.

$(SP)_{ch14}$ The metascientific applications of the two-level approach and the cognitive theory of metaphor cannot transgress certain *limits.* In particular, these limits result from

- fallacies,
- the peculiarities of plausible inferences,
- the fact that the potential of metascientific reflexion to contribute to the solution of objectscientific tasks is very restricted,
- the inherent shortcomings of the two approaches, and
- the fact that the case studies have to be regarded as thought expreriments.

$(SP)_{ch14}$ summarizes the second part of the solution to the main problem (Q‴) and thus supplements $(SP)_{ch10}$.

Nevertheless, the possible consequences of the limits enumerated are quite different. For example, as we have already mentioned, the status of plausible reasoning and fallacies is far from identical. In spite of their uncertainty, plausible inferences may be fruitful and progressive tools of scientific problem solving, while fallacies are not. To mention another example, the inherent shortcomings of particular cognitive semantic theories yield more serious limits than, say, the fact that metascientific reflexion can contribute to the solution of objectscientific problems only to a restricted extent. These examples boil down to the conclusion that a refined and sophisticated evaluation of the limits is needed which, however, would go far beyond the scope of the present work.

Chapter 15

Summary: The solution to the main problem

15.1 Looking back: The structure of the argumentation

- As a starting point, in the Introduction we raised the problem (Q):

  (Q) What are the prospects and limits of cognitive semantics?

- In Section 1.2 we replaced (Q) by (Q′) which we obtained by a possible explication of the term 'cognitive semantics' which includes both modularism and holism:

  (Q′) What are the prospects and limits of modular and holistic cognitive semantics?

- In Section 1.3, after surveying the state of the art in the philosophy of science, we put forward the thesis of the metascientific extension of cognitive semantics (MECS):

  (MECS) One of the tasks of cognitive semantics is to contribute to the solution of problems tackled by the naturalized philosophy of science in general and the cognitive science of science in particular.

  Since (MECS) specified one of the intended applications of cognitive semantics, it yielded, in turn, (Q″):

  (Q″) What are the prospects and limits of modular and holistic cognitive semantics *if* they are applied to scientific knowledge?

- In Chapter 2 we specified modularism and holism to paradigmatic examples: to the two-level approach as an example of modularism and the cognitive theory of metaphor as an example of holism. The generalized modularity hypothesis underlying the two-level approach said:

  (MH′) Human cognitive behaviour is organized in a modular way.

  The main hypothesis of the cognitive theory of metaphor was rendered in this way.

(HH′) Human cognitive behaviour is structured by metaphorical concepts.

In relating (MECS) to these two theories, we obtained the metascientific extension of the generalized modularity hypothesis (MMH′) and the metascientific extension of the main hypothesis of the cognitive theory of metaphor (MHH′), respectively:

(MMH′) Scientific knowledge is organized in a modular way along the lines of the generalized modularity hypothesis (MH′).

(MHH′) Scientific knowledge is structured by metaphorical concepts along the lines of the main hypothesis of the cognitive theory of metaphor (HH′).

Thus we obtained (Q‴) from the last two theses and (Q″):

(Q‴) What are the prospects and limits
a. of the metascientific extension of the main hypothesis of the cognitive theory of metaphor (MHH′),
and
b. of the metascientific extension of the generalized modularity hypothesis (MMH′)?

(Q‴) was regarded as the *main problem* of the book. In accordance with the line of argumentation sketched in the Introduction, in Parts II-IV (Q‴) was subdivided into more specific problems which were captured by particular *case studies.*

– Accordingly, Part II of the book was devoted to case studies illustrating certain *prospects* of cognitive semantics with respect to relevant aspects of scientific knowledge. In Chapter 4 (Q‴) was reduced to the following problem:

$(P)_{ch4}$ How can the following questions be answered, *if* the metascientific extension of the main hypothesis of the cognitive theory of metaphor (MHH′) as an example of holism is accepted:
a. What is the structure of explications in generative linguistics?
b. How is the structure of explications related to the structure of scientific explanations in generative linguistics?
c. To what extent are the answers to (a) and (b) related to semantic and pragmatic factors?

The case study carried out so as to exemplify the prospects of the metascientific extension of the two-level approach with respect to $(P)_{ch4}$ led to the following solution of $(P)_{ch4}$.

$(SP)_{ch4}$ If the metascientific extension of the main hypothesis of the cognitive theory of metaphor (MHH′) is accepted, then in generative linguistics

a. the relevant aspect of theoretical terms is that they are assumed to be metaphorical expressions which are manifestations of metaphorical concepts;
b. in accordance with (a) the explanans is identified with the source domain, the explanandum with the target domain, and there is a mapping between these two domains;
c. in accordance with (a) and (b) the cognitive theory of metaphor does not differentiate strictly between the semantic, the conceptual and the pragmatic aspects of theoretical terms and explanations, and it acknowledges, therefore, in this 'holistic' sense, the role of pragmatics in scientific concept formation.

– In Chapter 5 an analogous problem was raised:

$(P)_{ch5}$ How can the following questions be answered, *if* the metascientific extension of the generalized modularity hypothesis (MMH′) as an example of modularism is accepted:

a. What is the structure of explications in generative linguistics?
b. How is the structure of explications related to the structure of scientific explanations in generative linguistics?
c. To what extent are the answers to (a) and (b) related to semantic and pragmatic factors?

We obtained the following solution:

$(SP)_{ch5}$ If the metascientific extension of the generalized modularity hypothesis (MMH′) is accepted, then in generative linguistics

a. the relevant aspect of theoretical terms is that they are semantically underdetermined, because they rest on the relation between their semantic representation and their conceptual representation;
b. in accordance with (a), the relation between the explanans

and the explanandum is rooted in the particular mechanisms of conceptual selection which, in turn, is closely related to the particular mechanisms of conceptual shift and/or conceptual specification;

c. in accordance with (a) and (b) the assumption of a pragmatic module is implausible and central properties of theoretical terms should be explained not by using this category, but rather, by the specific interaction between relatively autonomous semantic and conceptual factors.

Nevertheless, although the two solutions to the problem of theoretical terms were analogous, due to the very different nature of the two approaches and the very different ways in which these solutions were obtained, we hypothesized that their further development would follow different directions and would reveal different prospects. So as to explore these prospects, it seemed to be plausible to show the differences in their possible extensions.

– Therefore, Part III centered around a highly topical problem in the philosophy of science which is currently the subject of great interest, namely, the question of what kind of relationship there is between the 'conceptual' and the 'social' aspects of scientific knowledge. Thus in Chapter 8 the prospects of the modular approach were tested with respect to reducing this problem to the following parts:

$(P1)_{ch8}$ What kind of relationship would there be between certain 'conceptual' and 'social' factors of generative linguistic theories, if the metascientific extension of the generalized modularity hypothesis (MMH′) were accepted?

$(P2)_{ch8}$ Can the two-level approach as an example of modular cognitive semantics be extended so that it captures the relationship between certain 'conceptual' and 'social' factors of scientific knowledge?

First of all, we introduced (IPS):

(IPS) The universal principles of the conceptual module underlying scientific knowledge are parametrized by the universal principles of the motivational module.

Then we obtained the solution to $(P2)_{ch8}$:

$(SP2)_{ch8}$ The two-level approach as an example of modular cognitive semantics is capable of being extended so that it captures the relationship between certain 'conceptual' and 'social' factors of scientific knowledge, because

a. it presupposes the two empirical hypotheses (MMH′) and (IPS), and
b. these, in turn, immediately yield the research strategy in (2).

Finally, by the help of the latter the solution to $(P1)_{ch8}$ was put forward:

$(SP1)_{ch8}$ If the metascientific extension of the generalized modularity hypothesis (MMH′) were accepted, the contribution of social factors to the conceptual structure of generative linguistic explanations would consist in the value of a parameter.

– We proceeded analogously with respect to the case study in Chapter 9:

$(P1)_{ch9}$ What kind of relationship would there be between certain 'conceptual' and 'social' factors of theories of AIDS, if the metascientific extension of the main hypothesis of the cognitive theory of metaphor were accepted?

$(P2)_{ch9}$ Can Lakoff and Johnson's cognitive theory of metaphor as an example of holistic cognitive semantics be extended so that it captures the relationship between certain 'conceptual' and 'social' factors of scientific knowledge?

The corresponding solutions were:

$(SP2)_{ch9}$ Lakoff and Johnson's approach as an example of holistic cognitive semantics is capable of being extended so that it captures the relationship between certain 'conceptual' and 'social' factors of scientific knowledge, because it can be supplemented by the methods of conversation analysis which are rooted in ethnomethodology.

$(SP1)_{ch9}$ If the metascientific extension of the main hypothesis of the cognitive theory of metaphor were accepted, then theories of AIDS would be structured by metaphorical concepts constructed in the course of verbal interaction.

– The task of Chapter 10 was to formulate basic generalizations which the case studies motivated and which concern the prospects of the metascientific application of cognitive semantics. The corresponding question was raised in this way:

$(P)_{ch10}$ By what further prospects can the findings of Chapters 4, 5, 8 and 9 be supplemented?

The answer to this question was as follows:

$(SP)_{ch10}$ a. The *prospects* of the cognitive theory of metaphor as an example of holistic cognitive semantics and of the two-level approach as an example of modular cognitive semantics result from the fact that they radically *widen* the scope of linguistic research. Among other things, they are capable of reformulating and solving classic problems and current quandaries in the philosophy of science.

b. Although the premises of the modular and the holistic approach are very different, both can be extended in a way which enables them to capture analogous problems of the philosophy of science. The findings they yield are original, fruitful and radically different from the perspectives of the analytic philosophy of science. In particular, both approaches are capable of accounting for central metascientific problems within their particular framework such as
   - the structure of theoretical terms,
   - the relationship between some 'conceptual' and 'social' aspects of scientific' knowledge,
   - the double-facedness of scientific knowledge,
   - the problem of the uniqueness of scientific knowledge,
   - the explication of the term 'theory',
   - the self-applicability of cognitive semantics,
   - the pragmatics of scientific discourse.

c. However, the considerations put forward within the two frameworks suggest very different solutions to these problems. They draw two different pictures of the way scientific knowledge in general and scientific concept formation in particular are structured and proceed.

$(SP)_{ch10}$ summarizes our main findings concerning the *prospects* of the metascientific extensions of modular and holistic cognitive semantics and constitutes *the first part of our solution to the main problem (Q''').*

– The central task of part IV was to exemplify some of the *limits* which the metascientific application of cognitive semantics cannot transgress. Chapter 12 raised a crucial problem of metascientific reflexion:

$(P)_{ch12}$ Can the cognitive science of science contribute to the solution of the objectscientific tasks of cognitive semantics?

The case study on the sceptical dilemma of cognitive semantics suggested $(SP)_{ch12}$ as an answer to $(P)_{ch12}$.

$(SP)_{ch12}$ a. The cognitive science of science and the practice of cognitive semantics are trivially *compatible* with each other.
b. Metascientific reflexion on the nature of cognitive semantics carried out by the methods of the cognitive science of science may *contribute immediately* to the solution of the objectscientific tasks of cognitive semantics.
c. At the same time the scope of this contribution is very *restricted* and metascience must not take over the tasks of objectscientific inquiry.

– In Chapter 13 the following problems were tackled.

$(P1)_{ch13}$ Could classic philosophical problems be reformulated and solved empirically, if the metascientific extension of the generalized modularity hypothesis (MMH′) were accepted?

$(P2)_{ch13}$ Could classic philosophical problems be reformulated and solved empirically, if the metascientific extension of the main hypothesis of the cognitive theory of metaphor (MHH′) were accepted?

The respective answers were:

$(SP1)_{ch13}$ As witnessed by a case study, the two-level approach was capable of solving a classic philosophical problem. At the same time, however, it cannot transgress certain limits. In particular, modular cognitive semantic argumentation is burdened, among other things, with the peculiarities of plausible inferences and the emergence of fallacies.

$(SP2)_{ch13}$ As witnessed by a case study, the cognitive theory of metaphor was capable of solving a classic philosophical problem. At the same time, however, it cannot transgress certain limits. In particular, holistic cognitive semantic argumentation is burdened, among other things, with the peculiarities of plausible inferences and the emergence of fallacies.

- In Chapter 14 we asked to following question:

$(P)_{ch14}$ By what further limits can the findings of Chapters 12 and 13 be supplemented?

Having reviewed some of the unsolved problems which were raised by the case studies in parts II and III, we obtained the following answer:

$(SP)_{ch14}$ The metascientific applications of the two-level approach and the cognitive theory of metaphor cannot transgress certain *limits*. In particular, these limits result from
- fallacies,
- the peculiarities of plausible inferences,
- the fact that the potential of metascientific reflexion to contribute to the solution of objectscientific tasks is very restricted,
- the inherent shortcomings of the two approaches, and
- the fact that the case studes have to be regarded as thought expreriments.

- Thus, we achieved the central aim of this book: *$(SP)_{ch10}$ and $(SP)_{ch14}$ yield the solution to the main problem (Q''')*. Figure 15.1 illustrates the structure of our argumentation.

15.2 Looking ahead: Perspectives for future research

Our aim in this study was neither to propagate cognitive semantics enthusiastically nor to refute it vehemently. If we had only emphasized the prospects, we would have reached to conclusion that cognitive semantics is a radically progressive contribution to the cognitive science of science. If we had concentrated on

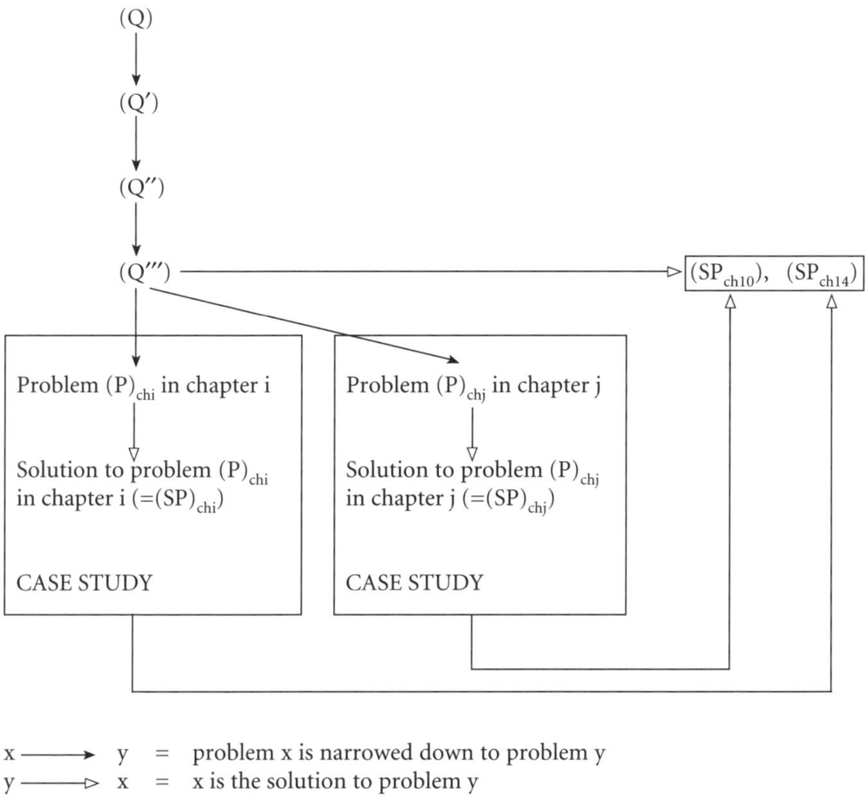

Figure 15.1

the limits, we would have carried the program of cognitive semantics to absurdity.

We pursued another strategy: put simply, that of trial and error. By applying two very different approaches to the investigation of certain aspects of the 'extreme' case of scientific knowledge, we wanted to reveal *both* their possible failure and their possible success. Accordingly, our main problem (Q‴) does not allow us to ask now, after the case studies have been carried out, if, on the basis of the prospects we discussed, cognitive semantics is a progressive enterprise or, rather, should be discarded because of its undeniable limits. The right question we have to ask so as to evaluate $(SP)_{ch10}$ and $(SP)_{ch14}$ is whether *both* the prospects and the limits motivate *fruitful problems to be tackled in future.*

$(SP)_{ch10}$ and $(SP)_{ch14}$ seem to suggest at least the following *new research topics for the cognitive science of science*:

- $(SP)_{ch10}$ and $(SP)_{ch14}$ are *the result of an attempt to widen the scope of the philosophy of science* by integrating cognitive semantic methods into the new field of the 'cognitive science of science'.
- One particularly important and instructive possibility to widen the scope of the philosophy of science to be emphasized here is its constructivity. Accordingly, one of the tasks for the future is the careful elaboration of the inventory by the help of which metascientific reflexion may contribute to the *improvement* of scientific problem solving.
- In Section 3 of the Introduction we outlined certain aspects of *thought experiments* without, however, discussing them in detail. Reflexion on the basic peculiarities of the case studies we carried out may therefore suggest useful *generalizations* concerning the way thought experiments in cognitive linguistics and in the cognitive science of science work.
- The argumentation carried out also serves to demonstrate how *plausible reasoning may work in cognitive semantics.* As Figure 15.1 illustrates, this applies both to the way we obtained the problems discussed and to the structure of the argumentation which led to the solutions proposed. Firstly, reducing one problem to another along the lines of Figure 15.1 (e.g. (Q) to (Q′), (Q′) to (Q″), (Q″) to (Q‴) and (Q‴) to the particular subproblems of the particular chapters) we made use of the heuristic method of *specification* discussed in the works of Polya. So as to refer to such a relation between the problems at issue we used the terms 'narrow down to' or 'specify to' in the same sense. Moreover, when arguing from the more specific hypotheses put forward in the case studies toward more general ones such as $(SP)_{ch10}$ and $(SP)_{ch14}$, we applied the heuristic method of *generalization.* Secondly, most (though not all) of our particular theses which we made use of in the course of the argumentation rested on plausible rather than deductively valid inferences. Nevertheless, at present we know very little of the way plausible inferences are structured and work in the course of our gaining knowledge of the world. Therefore, careful reflexion on the nature of *plausible reasoning* in cognitive semantics may significantly improve both cognitive semantic research in particular and the theory of plausible reasoning in general.
- The case studies also raised the embarrasing question of how to reflect on the possible *inconsistency* of models. Studying the way inconsistencies arise in cognitive semantic argumentation would be quite interesting from the point of view of current research into paraconsistency which is one of the new focal points both of the philosophy of science and logic.

Inseparably from these topics to be tackled by the cognitive science of science, $(SP)_{ch10}$ and $(SP)_{ch14}$ also suggest the need to think over some of the hidden problems of cognitive semantic research itself:

- $(SP)_{ch10}$ and $(SP)_{ch14}$ are the result of *an attempt to widen the scope of cognitive semantics* towards capturing some central aspects of scientific concept formation in particular and scientific knowledge in general.
- This leads to a rich inventory of new applications of cognitive semantics to analyse certain aspects of scientific knowledge which could be elaborated on in future. From the point of view of linguistics there is one particular aspect that deserves special attention: due to the self-reflexivity of cognitive semantics the metascientific application of *certain linguistic approaches* may be used to understand the nature of theory formation in *linguistics itself.*
- By the 'extreme' example of scientific knowledge we demonstrated that in cognitive semantics *case studies* which work as thought experiments may be applied, because they led to $(SP)_{ch10}$ and $(SP)_{ch14}$. In the *Introduction* we motivated the conscious use of thought experiments by the empirical weakness of cognitive semantic theories and the complexity of the object of investigation. Nevertheless, these two factors concern not only the particular metascientific applications of the two level-approach and the cognitive theory of metaphor, but raise the question of whether other applications of these two approaches should be regarded merely as thought experiments rather than real experiments. Other subfields of the subject matter of cognitive semantic theories are no less complex than scientific concept formation, and, as witnessed e.g. by Sections 2.2.1, 2.3.1, 14.2 and 14.3, the empirical uncertainty of the two approaches cannot be restricted to their metascientific application.
- $(SP)_{ch10}$ and $(SP)_{ch14}$ are the solution to (Q‴). Since (Q‴) is a specification of (Q), the question arises in what way and to what extent our findings may or may not be *generalized* so that eventually (Q) can be answered as well.
- Nevertheless, the possibility of widening the scope of both cognitive semantics and the philosophy of science encountered serious *limits.* Conscious reflexion on these limits is as important a task of cognitive linguistic research as exploring its capabilities.
- One of the most interesting results of our argumentation is that the two-level approach and the cognitive theory of metaphor "can lead to equally compelling, yet disparate, findings" (Haser 2003: 83; in the paper cited

> Haser's words apply to another situation, but they convincingly characterize our conclusion, too). The fact that "diversity in theoretical outlook" can "lead to thought-provoking different expositions of the same topic" (ibid.) refers to certain quite deeply rooted peculiarities of the state of the art in cognitive semantics which deserve to be thought over.

These implications of $(SP)_{ch10}$ and $(SP)_{ch14}$ outline *new research topics for the cognitive science of science.* They suggest that, despite the fallibility of the argumentation they rest on, further inquiry into hidden aspects of cognitive semantic research is *anything but unmotivated.*

Notes

Introduction

1. In Harder (1999) this statement is intended to characterize what we will call later 'holism'. Nevertheless, as we will see, it also applies to other trends within cognitive linguistics.

2. There are three reasons for the fact that we cannot avoid the pre-explicative treatment of our terminology. Firstly, as we will see in later chapters, both cognitive semantic theories and most current trends in the naturalized philosophy of science are to a considerable extent characterized by the vagueness of basic notions. Secondly, we will discuss several theories in this book; nevertheless, basic notions are handled differently in these theories and therefore it is not possible to define the terminology in a generally satisfactory way. Thirdly, our main concern is the examination of the prospects and limits of two cognitive semantic approoches, and this means that, in a certain sense, we are interested in how 'good' or how 'bad' the approaches at isseu are. Thus, in many cases we are justified in dispensing with the precise explication of the terminology we use, because "after all one can distinguish good from bad theories, or thoughts, or experiments without being able to define what exactly theories, thoughts or experiments are." (Peijnenburg and Atkinson 2003: 305)

3. Cf.: "The forms of *reasoning* used to manipulate the model will be the *same* as those we use to predict occurrences in the real world: inductive and deductive inferences are definitely permitted" (Cooper 1999: 267–268; emphasis added). Inductive inferences are often interpreted to include plausible reasoning, too. See e.g. Polya (1948), (1954) and Chapter 11.

4. Although this quotation mentions laws of nature, by analogy its implications also apply to those situations which will be the subject matter of the present investigations.

Chapter 1

1. In accordance with what we said in the Introduction, the terms 'empirical' and 'philosophical' will also be used pre-explicatively.

2. Cf.:

> I define cognitive science as a contemporary, *empirically based effort to answer long-standing epistemological questions* — particularly those concerned with the nature of knowledge, its components, its sources, its development, and its deployment. Though the term 'cognitive science' is sometimes extended to include all forms of knowledge — animate as well as inanimate, human as well as nonhuman

> — I apply the term chiefly to efforts to explain *human knowledge.* I am interested in whether *questions that intrigued our philosophical ancestors can be decisively answered, instructively reformulated, or permanently scuttled.* Today cognitive science holds the key whether they can be. […]
> A […] somewhat controversial feature is the claim that a key ingredient in contemporary cognitive science is the agenda of issues, and a set of concerns, which have long exercised *epistemologists in the Western philosophical tradition.* To my mind, it is virtually unthinkable that cognitive science would exist, let alone assume its current form, had there not been a *philosophical tradition* dating back to the time of the *Greeks.* (Gardner 1985: 6–7; emphasis added)

3. Cf.:

> At present most cognitive scientists are drawn from the ranks of specific disciplines — in particular, philosophy, psychology, artificial intelligence, *linguistics*, anthropology, and neuroscience (I shall refer to these disciplines severally as the 'cognitive sciences'). The hope is that some day the boundaries between these disciplines may become attenuated or perhaps disappear altogether, yielding a single, unified cognitive science. (Gardner 1985: 7; emphasis added)

4. The symposium mentioned in the quotation is not the only event that played a historical role in the birth of cognitive science. For example, in the same year another very famous meeting took place at Dartmouth College. See Gardner (1985) for a historical survey of these and other occasions.

5. More precisely, with respect to Government-Binding Theory.

6. Besides illustrating that both approaches can be subsumed under (3), the two quotations also indicate some of the basic differences between Chomsky's and Lakoff and Johnson's views.

7. Although in this way our argumentation reflects the general structure of definitions, the result of the following considerations must not be regarded as a satisfactory definition of the notion of 'cognitive linguistics', because, as mentioned in the *Introduction*, we cannot go into a comprehensive analysis of the state of the art; rather, we are compelled to restrict our considerations to certain factors which, from a heuristic point of view, seem to be relevant to (Q), without, however, drawing an unrealistic picture of the situation.

8. It is instructive to remark that in their introduction to the volume *Cognitive Linguistics: Foundations, Scope, and Methodology*, which is intended to give an overview of the methodological foundations and scope of the field, the editors give a similar account, restricting themselves to the enumeration of basic works instead of attempting a precise definition:

> Cognitive Linguistics established itself as a coherent, identifiable approach about a decade ago, marked by the first *International Cognitive Linguistics Conference* […] and by the founding of the journal *Cognitive Linguistics*, which first appeared in 1990. By that time, the major theoretical foundation had been laid and a substantial amount of empirical data had already been gathered to support and develop those theories, resulting in a number of scholarly publications that have since become widely cited classics: George Lakoff and Mark Johnsons's (1980) [=1980a]

> *Metaphors We Live By*, Leonard Talmy's (1983) 'How language structures space', Charles Fillmore's (1985) *Frames and the Semantics of Understanding*, Gilles Fauconnier's (1985) *Mental Spaces*, George Lakoff's (1987) *Women, Fire and Dangeorous Things*, Ronald Langacker's two volume work on the *Foundations of Cognitive Grammar* (1987, 1991 [=1991a]), Leonard Talmy's (1988) *Force Dynamics in Language and Cognition*, Brygida Rudzka-Ostyn's (1988) *Topics in Cognitive Linguistics*, George Lakoff and Mark Turner's (1989) *More than Cool Reason*, Eve Sweetser's (1990) *From Etymology to Pragmatics*, and the first volume in the *Cognitive Linguistics Research* series, Ronald Langacker's (1991) [1991b] *Concept, Image, and Symbol.* (Redeker and Janssen 1999: 1)

See also Verena Haser's review of this volume which is a concise evaluation of the state of the art in cognitive linguistics (Haser 2003).

9. Cf.:

> [..] Cognitive Linguistics is opposed to Saussurean and second-generation structuralist axioms, especially *dichotomies* such as *langue* vs. *parole*, synchrony vs. diachrony, syntax vs. semantics, lexis vs. grammar, etc. The claim of the arbitrariness of the linguistic sign is replaced by a search for *motivation* and the iconic principles of linguistic organization. [...] Cognitive Linguistics is opposed to *generative linguistics* [...], which sees language as an autonomous system, detached in principle from any other type of knowledge, especially *encyclopedic knowledge*. Cognitive Linguistics, in contrast, holds that there is no clear cut distinction between linguistic knowledge and encyclopedic knowledge [...] As Goldberg (1995: 5) puts it, 'knowledge of language is knowledge', just like any other type of knowledge, one can add. (Dirven 2000: 1; all emphases in the original).

10. Although much has happened in the decade since the publication of this quotation, the central idea according to which there are (at least) two main trends of cognitive linguistics, has remained — and this underlines the relevance of the dichotomy between holism and modularism. This may be witnessed by a more recent quotation by the same author:

> Cognitive theories of language and semantics by no means yield a homogeneous account of the state of the art. *Current research is shaped by a deep controversy centering on the positions of modularism and holism.* (Schwarz 2002: 279; my translation, A. K.; emphasis added)

Nevertheless, it must also be emphasized that 'cognitive linguistics' is considerably richer than the dichotomy between modularism and holism, and involves further trends. Nevertheless, it is still the case that the antagonism between modularism and holism decisively determines the state of the art.

11. For example, such very different theories as Jackendoff's, Lakoff and Johnson's, and Bierwisch and Lang's approach, each consider themselves to be 'cognitive/conceptual semantics', while they exclude other approaches from the domain of this term. Those whom we will call 'holists' later on do not allow the application of the terms 'cognitive linguistics' or 'cognitive semantics' to modular approaches and vice versa (see, for example, Gibbs 1996, Dirven 2000 etc.).

12. This does not mean, however, that we consider the interpretation (4) to be incorrect; rather, our choice is the result of a heuristic consideration motivated by the problem (Q) mentioned in the *Introduction*. Moreover, it should be admitted that the interpretation of 'cognitive linguistics' as (4) is more generally accepted than (5) which is, however, generally maintained, for example, in the German literature. This, nevertheless, is no argument against choosing, in the light of (Q), (4). Basically, the problem is a terminological and not a substantial one. The two trends to be compared are the same whether we choose to speak of two approaches within cognitive linguistics or of 'formalism' vs. 'functionalism' (as Langacker does) or, alternatively, 'first generation' vs. 'second generation' cognitive science in the sense of Lakoff and Johnson (1999).

13. There are several different ways of focussing on the theses enumerated here. For their partly similar, partly different presentation see e.g. Baldauf (1997), Kiefer (1995), Müller (1991), Taylor (1994), (1995a), (1995b). This enumeration of the basic theses rests on the literature mentioned.

It is important to remark that the criteria which yield the theses below are inhomogenous and integrate aspects of method, aims and the research object. Besides the fact that the literature just quoted does not clarify the criteria for delimiting these theses, in the present context there is a *substantial reason* for not differentiating between theses that concern the method, the subject matter and the aims of holistic and modular cognitive linguistics. For example, in Chapter 12 we will focus, among other things, on problems which arise *because* the proponents of certain approaches *cannot* distinguish between empirical hypotheses and methodological principles. Therefore, carrying out such a distinction at the outset would blur important problems and draw an incorrect picture of the state of the art.

14. See (11b) for the notion of 'module'.

15. The criteria on which these theses rest are in the same sense inhomogenous as those underlying (10a–h).

16. Let it be sufficient to mention just Schwarz (ed.)(1994) and Allwood and Gärdenfors (eds.)(1999) as paradigmatic surveys of cognitive semantics.

17. Again, this is a broad characterization of the term at issue rather than a precise definition. All the more so, because, among other things, we did not explicate the term 'meaning' which should be part of the definiens. At this stage of our argumentation a precise explication of this term is not possible, because the particular theories which are elements of the set of cognitive semantic theories, maintain very different views about the nature of 'meaning'.

18. It is worth remarking right away that, in full accordance with this conception of the subject matter of the philosophy of science, Chapters 8 and 9 will focus precisely on the problem of how the content of such "powerful theories which [...]" is shaped by conceptual and social factors.

19. Consider the following quotations taken from influential works:

> Epistemology is concerned with the foundations of science (Quine 1969a: 69);
>
> [...] the theory of knowledge (epistemology) has come to mean exclusively the methodology of the natural sciences and, more recently and belatedly, the social sciences [...] (Arbib and Hesse 1986: 1).

20. The retention of this distinction is *not* an arbitrary decision, but is closely related to the nature of the naturalized philosophy of science. See for example Stump's very important and clear argumentation:

> While it is certainly true that if we distinguish different levels of analysis in the sense that the science of science is metalevel discourse about the science rather than the object-level discourse of science, naturalism requires that the methodology of this metalevel discourse have the same status as the methodology of science. (Stump 1992: 457–460)

See also (14) below on the claim that metascience should have the same methodology as objectscience.

21. Just like the terms 'cognition' or 'conceptual' etc., 'empirical' is also used in a pre-explicative sense. In our context a satisfactory explication of this term is not necessary, although, of course, it is one of the central and most problematic notions of the philosophy of science.

22. See, for example, the quotations from Rouse's and Downes' writings in Section 1.3.3.

23. The structure of such an analogical inference may be reconstructed according to the following pattern:

1. Epistemology naturalized and the philosophy of science naturalized are similar.
2. Epistemology naturalized has the property X.

Therefore, the philosophy of science naturalized has the property X.

It is not irrelevant to remark that this kind of analogical inference is not our suggestion; rather, the fact that the naturalized philosophy of science rests on such analogical inferences is sufficiently witnessed in the literature. Also, acknowledging the use of analogical reasoning is in full accordance with our general strategy according to which we consciously make use of plausible inferences. See Chapters 11 and and 14 on this.

24. It is questionable whether we are justified in characterizing naturalized epistemology/the naturalized philosophy of science by a well-defined set of theses. Although the situation is very complicated and there have been many attempts to classify current trends, for the sake of argument here we have to content ourselves with a simplified picture. For more differentiated discussions of current naturalized epistemology see e.g. Almeder (1990), Maffie (1990), (1995), Kornblith (1985), Kitcher (1992), Koppelberg (1996), Rouse (1998) etc. The following survey is — in different contexts — also discussed in Kertész (2002b) and (2004).

25. Haack denotes this interpretation by SCIENCE (Haack 1993: 339). This corresponds roughly to German 'Wissenschaft'.

26. This corresponds to Haack's 'science'. Let us remark that it is not an easy task to define the distinction between natural science and other sorts of inquiry; therefore, it would be far beyond the scope of our considerations to attempt such a definition. This does not, however, undermine the consistency of our argumentation because in what follows nothing will depend on a precise delimitation of these areas of inquiry.

27. The independent origin of the cognitive turn and the naturalistic turn is also witnessed by the fact that by 'psychology' Quine meant behaviouristic and not cognitive psychology.

28. The following two textbook-entries reveal the same relationship between naturalism and the cognitive science of science.

> A cognitive approach to the study of science, then, is a species of naturalism in the philosophy of science [...]. It uses the cognitive sciences as a resource for understanding both the process of doing science and its products. (Giere 2001a: 41–42)
>
> A cognitive approach to science also represents a development of Quine's program, with his own behaviourism replaced by contemporary cognitivism [...]. (Giere 2001b: 309)

29. We will reflect on the nature of inferences applied in the course of our argumentation in Chapters 11, 13, 14 and 15.

30. In their book *Philosophy in the Flesh* Lakoff and Johnson reject Quinean epistemology / the philosophy of science naturalized:

> Cognitive science — the empirical study of the mind — calls upon us to create a new, empirically responsible philosophy, a philosophy consistent with empirical discoveries about the nature of the mind. This is not just old-fashioned philosophy 'naturalized' — making minor adjustments, but basically keeping the old philosophical superstructure. (Lakoff and Johnson 1999: 15.)

However, Lakoff and Johnson seem to have ignored the fact that naturalized epistemology / the philosophy of science came into contact with the cognitive turn and motivated the development of the cognitive science of science. Of course, as the quotation shows, *there is no contradiction beween the central assumptions of the cognitive science of science and Lakoff and Johnson's approach.* On the contrary: the latter may be interpreted as one methodological option within the former, admitting, of course, that there may be other possible approaches to the cognitive science of science as well. See also the remarks on the pluralism of the cognitive science of science in the next paragraphs. Thus the problem is not whether cognitive semantics is compatible with the cognitive science of science (although the latter is rooted in naturalism), but rather, what kind of relation there is between different and perhaps incompatible cognitive semantic approaches to the latter.

31. See for example the following generalization with respect to naturalized epistemology, which however, undoubtedly applies to the cognitive science of science as well:

> The term 'naturalized' epistemology derives from Quine: yet few philosophers even among those who apply this label to their own theories, are prepared to embrace naturalism in its 'replacement' version and abandon altogether the traditional issues of the theory of knowledge. (Shatz 1993: 117)

32. D. Stump summarizes this argument as follows with respect to the naturalized philosophy of science in general:

> A tension which has been ignored by the proponents of naturalized philosophy of science has been introduced into their program. On the one hand, naturalism demands unified method. On the other hand, naturalism also demands that the philosophy of science be true to science as practiced and, *pace* the positivists, science itself has shown not to be unified in its method. [...] While it is certainly true that we can distinguish different levels of analysis in the sense that the science

> of science is metalevel discourse of science, naturalism requires that the methodology of this metalevel discourse have the same status as the methodology of science. So, *if there are many methods of science, then there must be many methods of philosophy of science as well.* (Stump 1992: 459; emphasis added)

This argument also applies to the cognitive science of science in particular. In the light of what has been said in Section 1.3.2, in this quotation 'science' should be interpreted in the sense of (17)(a) or, alternatively, as SCIENCE in the sense of Haack (1993) (although Stump originally meant 'natural science').

Chapter 2

1. However, such an application was attempted in Kertész (1991), (1993).

2. Small capitals indicate *metaphorical concepts.* On this term see Section 2.3.1.

3. It is important to remark that all the central terms of the two-level approach are pre-explicative. The indeterminacy of the terminology and of basic hypotheses makes both its presentation and its application difficult.

4. For an explication of the structure of the argument which leads from (MH) to (MH′) below, see Chapter 13 of the present book.

5. The following simplified discussion of the theory will focus on its classic version as introduced in Bierwisch (1980), (1981), (1983a), (1983b), Lang (1994), Lang and Carstensen (1990), Lang et al. (1991) etc. Quotations will both illustrate and document our summary of the basic tenets. If they serve illustrative purposes only, so as to make the summary more understandable, they will appear in the notes. If they summarize important claims themselves, they will appear in the main text.

For the current version of the two-level approach see, for example, Bierwisch (1999a), (1999b). etc. It is worth remarking that the two-level approach has never been put forward systematically and its diverse exemplifications and applications differ both terminologically and in several other respects.

6. Cf.:

> Basically, all human cognitive behaviour is organized in a modular fashion. The structure formation underlying any concrete behaviour performance is *based upon* the *integration* of various relatively autonomous, task-specifically interacting systems and subsystems (MODULES). (Lang and Carstensen 1990: 6; emphasis added)
>
> The structure formation that underlies a *behaviour instance V* is the joint product of relatively autonomous, functionally interacting systems and subsystems. (Lang 1989: 266; emphasis added).
>
> Each of the individual structure-forming systems and subsystems comprises a distinct inventory of categorized ELEMENTS that are configurated to more complex representations SR, according to appropriate RULES which, in turn, are

> determined as to their format by both system-specific and general PRINCIPLES. (Lang and Carstensen 1990: 7; capitals in the original)

7. Cf.:

> A system *S* is autonomous to the extent that the representations *SR*, determined by *S*, are determined by specific principles which are only valid for *S*.
> Two systems S_1 and S_2 interact to the extent that the representations SR_1, determined by S_1, contain *parameters* which are instantiated by appropriate values from the representations SR_2, determined by S_2. (Lang and Carstensen 1990: 7; italics added)

8. The precise meaning of the verb 'depend' in these claims is to be left open. In fact, it may differ from case to case. One extreme is that the value of P_1 is identical with the value of P_2. The other extreme is that the value of P_2 merely indirectly implies a set of properties which may set the parameter P_1. Between the two extremes there is a large range of possibilities. Later on we shall illustrate how the idea of parametrization works in the case of scientific knowledge.

9. The brackets '<' and '>' denote an ordered n-tuple.

10. The context *ct* contains at least a possible world, but the question of what further information *ct* includes is left open. With respect to *ct* the move from formal semantics to modular cognitive semantics may be exemplified by the following argumentation.

> According to possible worlds semantics, the context *ct* and the utterance meaning *m* would have to be specified in terms of possible worlds and objects, which are characterized by means of certain set theoretical structures. [...] *these are to be interpreted as the formal structure of mental states*, i.e. of *internal representations* of worlds, objects, states of affairs, etc. Instead of being just arbitrary set theoretical structures, possible worlds are then generated by the system [...] of perceptual and conceptual rules, schemes, and operations of a person *p* [...]. (Bierwisch 1980: 6; emphasis added)

11. Cf.:

> [...] the linguistic and the communicative aspect are determined by different and largely independent principles and rules even in clearcut cases of verbal communication, i.e. speech acts. In other words, language and communication (or more generally: *social interaction*) are based on different systems of knowledge. [...] Given a domain of language use based on the rules and structure of language, and a domain of communication based on its own principles, rules, and structures belonging to the more inclusive area of social interaction, speech acts must be construed as instances interconnecting the two domains in a particular way. To put it in simple terms: A speech act makes a linguistic utterance, mainly by virtue of its meaning, the bearer of what would best be called a communicative sense. Notice that a communicative sense belongs to the domain of *social interaction* and can in general be implemented in various ways, among which the use of verbal utterances is the most elaborate and often the most effective one. (Bierwisch 1980: 3; emphasis added)

12. Since these notions seem to be, at first sight, fairly abstract, it may be useful to illustrate them by a simple example. Say Peter, a teacher of literature, says the following to his pupils during a class devoted to European poetry:

(i) *The great Hungarian poet Petőfi can tell young people a lot about love and freedom.*

The utterance *u* of (i) counts as an *instance of behaviour* which is assumed to consist of a set of representations. For example, Peter's gestures which accompany his utterance, the way he operates his organs of articulation are assumed to be represented in his mind *motorically*. The sentence which he utters is represented in the *grammatical module* and is, in turn, subdivided into a phonetic, a morphosyntactic and a semantic representation *pt, syn* and *sem*, respectively. The *phonetic representation* consists of a sequence of segments together with suprasegmental properties like accent, intonation and syllable structure. The *morphosyntactic representation* includes, among other things, the information that *can* and *tell* belong to the same VP. Moreover, the different contexts *ct* in which (i) may be embedded, result in different conceptual representations of the verb *tell*. Depending on these, (i) yields, for example, the following utterance meanings *m* (i.e. conceptual interpretations):

(ii) a. *Petőfi <u>delivered utterances</u> from which young people can learn a lot about love and freedom.*

b. *Petőfi <u>wrote poems</u> from which young people can learn a lot about love and freedom.*

c. *Petőfi <u>lived and died in a way</u> from which young people can learn a lot about love and freedom.*

These utterance meanings — associated with the utterance *u* and the context *ct* — result in different meaningful utterances. If one of the meaningful utterances is given, Peter may carry out different speech acts *sa* with respect to particular interactional settings *ias* — say, when in the classroom he wants pupils to understand how to consolidate personal life with abstract moral values and historical actions. (Petőfi was not only a great poet, but also one of the heroes of the Hungarian revolution in 1848–49. He died in battle in 1849.) Depending on the particular properties of *ias*, the meaningful utterance (ii)(c) may be a piece of advice, a warning, a proposal etc.

13. Cf.:

> [...] the conception of language as a *mental system* is not at variance with the fact that language is a *social* phenomenon. It rather emerges and functions as a constitutive factor of social organization just because it is an intrinsic capacity of the human mind. Only on this basis can it be shaped by and act upon social conditions in ways that are to be studied as a genuine question on its own. (Bierwisch 1983a: 34; emphasis added)

14. Here 'ambiguity' means that a given term may have several interpretations represented in the lexicon and these interpretations differ from each other semantically. 'Vagueness' means that we cannot decide whether an object is in the extension of a certain predicate or not. For surveys of how the two-level approach treats polysemy see e.g. Pethő (2001a), (2001b).

15. An example of how this works will be mentioned in (14). Nevertheless, the way lexical items are decomposed will play no role in the argumentation of this book. Therefore, here we dispense with going into a detailed explanation of the mechanisms of relating componential semantics to intensional semantics.

16. Cf.:

> Lexical items, even if they are not ambiguous in any ordinary sense, are in general related not to single concepts, but rather to systematically connected *families of concepts*, each of which may become, depending on the context of interpretation, its conceptual interpretation. The mechanisms involved are *conceptual shift* and *conceptual specification.* (Bierwisch 1983a: 66; emphasis added).

17. 'Conceptual domain' is a frequently used notion of two-level semantics; however, as most notions, it is used pre-explicatively and we cannot explicate it, either.

18. A lengthy discussion of the formalism exemplified here would be beside the point. It has already been mentioned in (10) that intensional semantics is combined with componential analysis. Although familiarity with these techniques should be presupposed, the argumentation of the present book will be understandable even if we do not give an overview of intensional semantics and componential analysis here. Let it be sufficient to quote Bierwisch with respect to the lambda operator:

> I am using [...] the lambda operator in the usual way, i.e. as an abstractor binding a variable such that if x is a variable of category a and φ any (basic or complex) element of category b, then $\lambda x\varphi$ is a complex element of category b/a. From this the usual principles of lambda conversion follow immediately, i.e. $\lambda x(Px) = P$, hence $Pa = [\lambda x(Px)]\ a$. Bierwisch (1983a: 54)

19. Here 'variability' means the same as 'semantic underdetermination'. See Bierwisch (1983b).

20. Of course, much depends on the proper explication of the notions 'litereral meaning', 'non-literal meaning', 'neutral context', 'non-neutral context', 'nearest neutral context'. In spite of this, the two-level approach treats these notions pre-explicatively. As we will see later, this does not undermine the applicability of the two-level approach to scientific concept formation.

21. As we will see in Chapter 5, we find such specific circumstances in theoretical contexts where the combination of metaphorization with the very different principles of conceptual shift, conceptual specification and conceptual selection play an important role with respect to scientific concept formation.

22. This integrative nature of the two-level approach legitimizes its integration, among other things, with the sociological approach we will suggest in Chapter 8.

23. The structure of this inference is crucial for the evaluation of our later findings. See Section 13.2.2 on this.

24. Although 'scientific knowledge' is a central term of current philosophy of science both in the sociological and the cognitive tradition, it is used pre-explicatively. Therefore, in accordance with our general strategy, we will not explicate it, either. Some consequences of the terminological vagueness which we cannot avoid will be discussed in Chapter 14.

25. The terminology and the main theses of the cognitive theory of metaphor are very vague, and this vagueness will be reflected in the presentation, too.

26. This is not intended to be a comprehensive discussion of the cognitive theory of metaphor. The survey will rest mainly on its classic version as introduced, applied and further developed e.g. in Johnson (1987), Lakoff (1987), Lakoff and Johnson (1980a), (1999), Kövecses (1986), (1990), (2000) etc. In the present context — just as in Chapter 1 — we focus on a relatively simple introduction of some basic tenets of the theory which are central to our concerns. Besides the works mentioned, in the following survey we will also make use of their discussions in Baldauf (1987) and Jäkel (1987). Further aspects will be elaborated on in later chapters in connection with the problems to be raised there.

27. This is what the *Ubiquity Hypothesis* amounts to.

28. Cf.:

> Metaphor is fundamentally conceptual, not linguistic, in nature. Metaphorical language is a *surface manifestation* of conceptual metaphor. (Lakoff 1993: 244; emphasis added).

29. The key idea is this:

> A metaphor is not merely a linguistic expression (a form of words) used for artistic or rhetorical purposes; instead, it is a process of human understanding by which we achieve meaningful *experience* that we can make sense of. A metaphor, in this 'experiential' sense, is a process by which we understand and structure *one* domain in terms of *another* domain of a different kind. (Johnson 1987: 15; emphasis by underlining in the original).

30. This interpretation of the cognitive theory of metaphor yields a very important consequence with respect to its metascientific application: as we will see in Chapter 4, it facilitates, along the lines of naturalism in general and the cognitive science of science in particular, the *metascientific explanation* of certain aspects of objectscientific research.

31. This assumption is called the *Invariance Hypothesis* which has given rise to heated discussions in the literature.

32. See Chapter 4 on this.

33. Jäkel intends to generalize what Lakoff and Johnson call 'higlighting and hiding'.

34. In Lakoff (1987), Lakoff and Johnson (1980a), (1999) there are extensive and lengthy discussions of this contrast which we presented here in a very simplified manner. In Lakoff and Johnson (1980a) and Lakoff (1987), for example, the dichotomy 'objectivism vs. experientialism' is made use of, whereas in Lakoff and Johnson (1999) the dichotomy 'first generation cognitive science vs. second generation cognitive science' is preferred:

> What we are calling 'first-generation' versus 'second generation' cognitive science has nothing to do with the age of any individual or when one happened to enter the field. The distinction could just as well be called 'disembodied' versus 'embodied' or 'assuming tenets of formalist analytic philosophy' versus 'not assuming tenets of formalist analytic philosophy'. The distinction is one of philosophical and methodological assumptions. (Lakoff and Johnson 1999: 78)

We will preserve the label 'cognitive theory of metaphor' instead of the alternatives 'second generation cognitive science' or 'experientialism' so as to indicate that in the present study the assumed metaphorical structure of scientific concept formation will be focussed on.

Chapter 3

1. Before doing so, however, a terminological remark is in order. In the following summary we *do not* differentiate between natural science and 'science' in the sense introduced in (17) in Chapter 1. The reason for this decision is as follows. Firstly, the analytic philosophy of science raised the problem of explanations, explications and theoretical terms with respect to natural science. Secondly, these problems arise with respect to other fields of 'empirical' inquiry (in whatever sense) as well. An instructive example is *generative grammar*, which claims to proceed like the natural sciences do — especially as regards the construction and use of theoretical terms, scientific explanations and explications —, although it is fiercely disputed whether it really does so. (See also the heated debates of the seventies concerning this issue as summarized e.g. in Cohen (ed.) (1974), Cohen and Wirth (eds.) (1975), Itkonen (1978), (1983), Kertész (1991), Perry (ed.) (1980), Wirth (ed.)(1976), Wunderlich (ed.) (1976) etc.) Since the case studies we want to develop will relate the problems of explanation, explication and theoretical terms to generative grammar, we may treat these in a similar way as they are treated in the natural sciences. Therefore, in outlining our starting point it is fully acceptable that we simply summarize basic views about theoretical terms with respect to natural science and then, in the course of the case studies, we will examine whether this view applies to generative grammar as well.

Of course, one of the questions we cannot raise here is how to define 'natural science' and how to delimit it from other kinds of inquiry. We have to leave this question open, all the more so, because in recent times the traditional assumptions of the unity of natural scientific method have been clearly undermined. See e.g. Dupré (1982), (1993).

2. Since it was physical theories which were considered as the paradigmatic examples of theory formation, 'measurement' played a central role.

3. The literature is very rich, so the following references are nothing more than representative examples and cannot stand for a comprehensive survey of the state of the art.

4. See Putnam (1962) and Suppe (1977) for the notion 'received view'. It was introduced to refer to a system of assumptions which most trends within the analytic philosophy of science share, despite their differences.

5. It is assumed that laws can be explained as well, but here we don't need to go into this, because in our later argumentation we will not touch on the explanation of laws.

6. Let us emphasize again that in selecting these properties as part of the 'received view' of explanations we want neither to blur the problems mentioned nor to claim that there is a unified treatment of explanations.

7. 'Exactness' and 'precision' mean that the vagueness and the ambiguity of terms are excluded by the use of some formal language.

8. 'Hypothetical construct' is another name for 'theoretical term'.

9. By 'deduction' valid logical inference is meant.

10. Nevertheless, there are well-known attempts to develop 'pragmatic' accounts of scientific explanations. See e.g. Stegmüller (1983), Kertész (1988).

Chapter 4

1. Jäkel (1997), Liebert (1997a, b etc.) and Drewer (2003) are systematic applications of the cognitive theory of metaphor to scientific knowledge. In this respect our analyses are not new but rather, fit into current research. We will refer to these works; a comprehensive overview of them would be beside the point. Our findings will be in accordance with their suggestions.

2. The examples enumerated below are freely adapted from Riley (1987) and the basic literature of generative linguistics. These examples serve illustrative purposes within this case study only, and are not meant to be the result of a comprehensive and complete analysis of the terminology of generative linguistics.

3. One may argue that these theoretical terms express well defined structural relations and could, therefore, be replaced for example by symbols like *P, Q, R* etc. which serve the very same tasks. However, the point is that *if* we assume the metascientific extension of the main hypothesis of the cognitive theory of metaphor (MHH'), then this must not be the case: from Lakoff and Johnson's approach it follows immediately that it is the metaphorical nature of theoretical terms which shapes the cognitive structure of the theory and that the cognitive structure of the theory would be different if metaphorical expressions were replaced by something else.

4. Please note that our formulations are very vague. At this point of our argumentation this is not a shortcoming of the present analysis, but a straightforward consequence of the cognitive theory of metaphor and the "received view". The question as to what extent this vagueness indicates a limit of the metascientific application of the cognitive theory of metaphor will be raised explicitly in Chapter 14.

Chapter 5

1. Here again we must not forget that we do not claim that theoretical terms in generative linguistics do indeed work along the lines of the metascientific extension of the two-level approach to cognitives semantics. Rather, we proceed *hypothetically*: we set out to explore what *would be the case if* (MMH') were acceptable. See also Chapter 14 on the methodological consequences of this strategy.

2. Due to (MMH'), the context *ct* includes pieces of information within the theory which make up the conceptual environment of the given term. Among other things, this is what the first quotation in Section 2.1. suggests.

3. However, see the problems raised in Chapter 14.

4. These conceptual representations may be terminologically distinguished as emphasized in Bierwisch 1983a: 67; also quoted in Section 2.1.

5. In the above quotation the term 'precision' and 'imprecision' witness that Tsiapera accepts the 'received view'. This does not undermine our argument, however. This simply means that if we accept that theoretical terms may be treated as literal expression, the two-level approach will be entitled to handle their literal meaning with its own means. These means, in turn, suggest that it is semantic underspecification which matters and not the idea of 'precision' advocated by the analytic philosophy of science.

6. This is neither an exhaustive enumeration of the intrpretations of *command* nor do we strive to present a detailed analysis here. The following considerations serve only illustrative purposes. See also Kertész (1991).

7. Different members of the family may be distinguished terminologically. It may be an interesting task to analyze the members of this family with respect to the question of when a certain member is distinguished from another terminologically and when not. Nevertheless, this question is not our concern in the present context of argumentation, because we have to focus on the relationships between the common semantic representation of all members of the family and their different conceptual representations.

8. 'Refer' is used pre-explicatively and must not raise the general problem of 'reference' here.

9. See, among other things, Elffers (2003), and the literature cited there.

10. See also the considerations in Hesse (1980), which are analogous to these conclusions although they concern the role of metaphors in scientific explanations.

11. These considerations will be continued in Section 10.2.5.

Chapter 7

1. We will refer to the second edition of Kuhns work which contains important additions to the first edition, but we must not forget that the first edition came out in 1962 and by the end of the sixties it had already become one of the most discussed works of the philosophy and history of science.

2. Theses tenets will be explained in Section 8.1.2.

3. For a discussion of some of these works see Kertész (1993).

4. See vols. 19(1989)-22(1992) of *Social Studies of Science* and a series of later publications. Kertész (1993) is devoted to a detailed analysis of the debate.

Chapter 8

1. See also Fuller (1993b) for an evaluation of the role generative linguistics played in this controversy.

2. In the debate between the proponents of the cognitive science of science and the sociologists of knowledge the basic terms were used vaguely, and the terminology is not unified either. So, expressions like 'cognitive', 'mental', 'psychological', and 'conceptual' factors of scientific knowledge are meant to refer roughly to the same sort of factors. Therefore, here we have to chose between these terms. Owing to the basic assumptions of the two-level approach which were discussed in (1) in Section 2.2.1, we will choose the term 'conceptual' for obvious reasons to be explained in Section 8.2.1.

3. Here again 'empirical' is used pre-explicatively.

4. Although Fuller uses the term 'naturalized epistemology', on the basis of what we said in Section 1.3.2 this applies to the naturalized philosophy of science as well. See also Downes (1993), Maffie (1995), Shatz (1993).

5. The present chapter makes use of findings which were also discussed in Kertész (1993).

6. It is worth emphasizing again that these modules are hypothetical constructs whose existence is only assumed but has not so far been proved by empirical evidence (see Bierwisch and Lang 1989 on this). See also Section 14.3 on this problem.

7. So as to avoid confusion, we will distinguish between the terms 'metascience' and 'objectscience' so that at every stage of our analysis the reader knows what kind of explanation, or description, or instance of behaviour etc. we are talking about. This distinction is taken for granted in the philosophy of science although it is highly problematic. Therefere, it is important to emphasize that this is only a terminological distinction which has no ontological consequences. All the more so, because, by assumption, instances of object- and metascientific behaviour rest on the interaction of the same modules. See also the considerations in Section 10.2.

8. This means that a theory is treated as an instance of behaviour which consists, among other things, of a conceptual, a grammatical, an interactional and a motivational representation. All of these aspects are decisive in the constitution of theories, and it is these particular representations which require a metascientific description and/or explanation. Metascientific explanations of objectscientific theories will therefore differ according to the particular choice of the kind of representation to be explained. For example, the metascientific explanation of a conceptual representation will naturally be very different from the explanation of its motivational representation. See also Section 10.2.3.

9. In Kertész (1991) I assumed that it is the context which is a free parameter associated with conceptual principles (see (11) in Section 2.2.1 and Section 8.3.4) and that there are at least three parameters associated with the different motivational principles: namely, 'social interests', 'the degree of the inner organisation of a group' and 'the strength of group boundaries', all of which are assumed to be possible candidates for motivational parameters.

10. A clear differentiation between rule-oriented and principle-oriented explanations in generative grammar, along with a careful analysis of the historical and social context in which they have been developed, is Newmeyer (1991). See also a simple contrasting of these two perspectives in the quotation from Chomsky (1991) in Chapter 4.

11. See the discussion between Nola (1992) and Bloor (1992).

12. Let it be sufficient to refer to Kertész (1991) for further considerations.

13. See, by way of illustration, Riley on this point. After comparing the theoretical terms of the Standard Theory with those of Government-Binding Theory, she comes to the following conclusion:

> [...] researchers working within each theory have been guided by different *priorities and goals*, and these in turn are reflected in the jargon associated with each theory. (Riley 1987: 178; emphasis added).

The following argumentation is the immediate continuation of the findings of the case study in Chapter 5.

14. We know that all constituents of modules are treated as mental representations and this facilitates the interaction between them. Therefore, there is nothing awkward in assuming that the value of a motivational parameter is adopted in some way or other by a conceptual parameter, because there is no qualitative difference between the two values — both are assumed to be represented in the mind.

Chapter 9

1. Please note that our argumentation will be as hypothetical as in Chapters 4, 5 and 8. In fact, $(P1)_{ch9}$ and $(P2)_{ch9}$ amount to the following *What would be the case if ...?* question: "What kind of relationship would there be between certain 'conceptual' and 'social' factors of AIDS research, *if* the cognitive theory of metaphor could be extended so that it captures the relationship between certain 'conceptual' and 'social' aspects of theories?" Nevertheless, in analogy to how we proceeded in the previous chapter, it is from an argumentative point of view important to split up this question into $(P1)_{ch9}$ and $(P2)_{ch9}$ and consider them separately.

2. Nevertheless, the scope of the present study does not permit an analysis of the compatibility of Latour and Woolgar's approach with the conversation analytic investigation which will be put forward below. Let it be sufficient to make two remarks. Firstly, the basic stance of Latour and Woolgar's approach concerning the nature of scientific terms is in full accordance with the implications of Chapter 4 and the analyses to be carried out in the present chapter. The main issue is that the basic aspects of scientific concept formation are assumed to rest on the same mechanism, whatever these are, as those of everyday concept formation:

> By this introduction to the microprocesses of the fact production we have tried to show that a close inspection of laboratory life provides a useful means of *tackling problems usually taken up by epistemologists*; that the analysis of these micro-

> processes *does not in any way require the a priori acceptance of any special character of a scientific activity* [...] (Latour and Woolgar 1979: 183; emphasis added)
>
> [...] scientific expressions are no better able to yield a determinacy of meaning than any employed in 'nonscientific' or common sense contexts. (Latour and Woolgar 1979: 184)

Secondly, although Latour and Woolgar's investigations are — for practical and technical reasons mentioned in their book — not strictly conversation analytical, they clearly consider the application of conversation analysis as one of the intended methods of the sociology of knowledge which is fully compatible with their approach:

> Tape recorders could not be used, so these notes lack the precision necessary for 'conversation analysis'. Even in their somewhat crude or 'tidied' state, however, these discussion notes provide a useful opportunity for a close analysis of the construction of facts. (Latour and Woolgar 1979: 154)
>
> A full analysis, particularly one that aspired to the rigour of 'conversation analysis' [...], would demand much more detailed treatment than given here. (Latour and Woolgar 1979: 184)

3. Of course, there is no such thing as conversation analysis; rather, there are numerous methods which differ from each other in several respects but which can all be traced back to ethomethodology. Therefore, this summary is an oversimplification rather than a realistic account of the state of the art in conversation analysis. Nevertheless, it will be sufficient for the present purposes. For those readers who are well informed, it will be clear that what we can do is nothing more than to outline some aspects which most trends within ethnomethodological conversation analysis share on a relatively high level of abstraction.

4. The transcriptions were based on Gutfleisch-Rieck et al. (1989). Some of the central conventions include the following (see Liebert 1997a: 160):

arrow up/down	intonation up/down
capital letters	stressed syllables or other units
underlined words	overlapping turns of two or more speakers
:	acoustic expansion
*	pause of 1 to 1.5 seconds
/	re-starting the phrase after making a mistake
(CAPITAL LETTERS) #...#	Paralinguistic feature (for example LAUGHTER) followed by its range

5. Scientific speakers are abbreviated with 'S + Number'.

6. In this respect the classic work is Hesse (1966).

7. The notion 'ICM' was introduced in (28) in Section 2.3.1.

8. The thesis of autonomy is discussed in detail in Finke (1982) with respect to linguistics and the theory of literature.

9. This notion of constructivity will be elucidated in Section 11.2.

10. For a detailed epistemological discussion of this reserch strategy see Kertész (2002b) and (2004).

11. Cf.:

> The many aspects which are involved in the investigation of metaphors in verbal interaction deem it necessary to consider approaches from a number of different disciplines. These include central concepts of cognitive and social psychology (problemsolving, analogical reasoning, perspective), conversational analysis (turn-taking, aspect use), and cognitive linguistics (levels of metaphor models). (Liebert 1997a: 150)

Chapter 10

1. Of course, linguistics is not the only reflexive discipline. See, for example, Bloor's thesis of the reflexivity of sociology quoted in Section 8.1.2. Or think of further possible subdisciplines of cognitive science which may be self-reflexive, too.

2. See Kertész (2001b) on some general epistemological and textological aspects of the 'pragmatics of scientific discourse'. The papers in Kertész (ed.)(2001) outline different approaches to the latter.

3. See the recent discussion on metaphor identification in vol. 11 (2002) of *Language and Literature.*

4. As already mentioned in Chapter 6, since the two metascientific approaches use different terms and different background assumptions which yield different interpretations of the issues they touch on, the problems enumerated are only analogous but not identical.

Chapter 11

1. See e.g. Laudan's 'normative naturalism' as a paradigmatic example of a systematic solution. For alternative views see e.g. Goldman (1986), Kornblith (1985), (1993), Stich (1990) etc.

2. This will be exemplified in Chapter 13.

3. As regards modern literature on fallacies, see e.g. Hamblin (1970), van Eemeren and Grootendorst (1992), Walton (1987), (1997), Woods and Walton (1990) etc.

4. It would be beside the point to enumerate the classic fallacies and comment on them here. In Chapter 13 two of them will be important, namely *circulus vitiosus* and *regressus ad infinitum.* Their structure and relevance will be clear from the considerations presented there.

Chapter 12

1. The relevant literature of the debate includes, among others, Boas (1984), Botha (1981), Carr (1990), Cohen (ed.)(1974), Cohen and Wirth (eds.)(1975), Finke (1979), Itkonen (1978), (1983), Kertész (1991), Oesterreicher (1979), Perry (ed.)(1980), Wunderlich (1976) etc.

2. See e.g. Kiefer (1995), Müller (1991), Schwarz (1992), (1997), (2002), Taylor (1994), (1995a), (1995b) etc.

3. Let us remember that the notion 'empirical' is used in a pre-explicative sense. In the context of the present book nothing depends on a precise explication which would, due to the well-known difficulties involved in this term, lead to an unnecessary complication in our line of argumentation. The decision that the term 'empirical' will not be explicated here is thus the result of heuristic considerations motivated by the nature of the problems the book focuses on.

4. However, the weak version is unsatisfactory, too:

> As Murphy (1996, 1997) argued, there is no evidence to support the strong or weak versions of the metaphoric representation claim. However, an even weaker version of this claim might merit consideration. On this version, metaphors such as THEORIES ARE BUILDINGS do not structure our understanding of theories in general (the strong version), nor do they exert an indirect influence on the structure of our knowledge of theories (the weak version). Nevertheless, they are part of our knowledge of how people talk about such abstract concepts and can play an important role in our understanding of figurative expressions that refer to such concepts. In the field of psycholinguistics, Raymond Gibbs has been the major proponent of this version of the representation claim, which amounts to a process claim — that conceptual metaphors underlie the processes with which we interpret figurative language. (McGlone 2001: 98)

Nevertheless, we do not have to go into this more refined interpretation of the main hypothesis of the cognitive theory of metaphor, because already the weak version avoids the metaphorical circle.

5. For example, this may mean that, other things being equal, a Jackendoff-type semantics may be more plausible than a radically holistic or modular one.

Chapter 13

1. Cf. the quotations in Section 1.2.1 which illustrate the subsumption of Chomsky's generative grammar and Lakoff and Johnson's cognitive semantics under the linguistic subdiscipline of cognitive science. The argumentation put forward in the present chapter is closely connected to the content of these quotations.

2. This is underlined not only by Gardner who is one of the most influential historiographers of cognitive science but also by Chomsky's opponents such as George Lakoff and Mark Johnson (see Lakoff and Johnson 1999: 75).

3. At this point we have to leave open the question as to what is meant by the 'empirical' nature of generative linguistics; therefore, the term 'empirical' will be used pre-explicatively. However, there is no doubt that Chomsky developed his theory by trying to follow the values of 'empirical' natural science in general and physical theories in particular. All proponents of generative linguistics consider their field of research to be 'empirical', whatever this means.

4. This argument corresponds to the following kind of plausible reasoning:

> If there are two different conjectures, *A* and *B*, aimed at explaining the same phenomenon, we regard them as opposed to each other even if they are not proved to be logically incompatible. These conjectures *A* and *B* may or may not be incompatible, but one of them tends to render the other superfluous. This is enough opposition, and we regard *A* and *B* as *rival* conjectures. There are cases in which we treat rival conjectures almost *as if* they were incompatible. For example, we have two rival conjectures *A* and *B* but, in spite of some effort, we cannot think of a third conjecture explaining the same phenomenon; then each of the two conjectures *A* and *B* is the 'unique obvious rival' of the other. [...] In general, the relation between rival conjectures is similar to the relation between rivals in any other kind of competition. If you compete for a prize, the weakening of the position of any of your rivals means some strengthening of your position. [...] If you have a unique obvious rival, any weakening or strengthening of his position influences your position appreciably. (Polya 1954: 31)

5. Let us remark for the sake of clarity that the last sentence of the quotation applies not only to the two Modes mentioned but to the others as well.

6. The role which the problem of the Criterion played in ancient philosophy is characterized by A. A. Long like this:

> At the time of Ptolemy and Sextus it had become virtually de rigeur for any thinker to state his position on the 'criterion of truth' [...]. (Long 1989: 154)

7. For useful overviews of these and further manifestations of the problem of the Criterion see e.g. Floridi (1993), (1994), (1996), Sas (1999), Striker (1990), Huby and Neal (1989), etc.

8. For a considerably more complex reconstruction see Floridi (1996).

9. It is important to remind the reader that current trends in epistemology often focus on problems concerning the nature of *scientific* knowledge. For example, the first sentence of Quine's seminal paper, which led to his programme of naturalized epistemology, is the following:

> Epistemology is concerned with the foundations of science. (Quine 1969a: 69)

10. See, for example, Chomsky (1986), Fodor (1983), Jackendoff (1983), (1990), Bierwisch and Lang (eds.)(1989), etc.

11. x = scientific knowledge, y = human behaviour, P = modularly organized.

12. It is worth mentioning that this is analogous to Chomsky's solution to Plato's problem.

13. Here, as already mentioned, the *capability* of the two-level approach to solve these problems is stressed, because we did not go into the details of the solution.

14. Nevertheless, in Kertész (2004) I proposed an epistemological framework in which the problem of such a regress can be avoided. See also Kertész (2002b) for a concise summary of this framework.

15. Therefore, our discussion of $(P2')_{ch13}$ will be considerably shorter than that of $(P1')_{ch13}$ and will use quotations from and references to Lakoff and Johnson (1999).

16. See for example Neisser (2001) for a fierce criticism of Lakoff and Johnson (1999).

17. See Kertész (2004) and (2002b) for a framework which attempts to handle this problem.

18. For example, linguistic theories as analysed in Chapter 8 or theories of AIDS as shown in Chapter 9.

Chapter 14

1. These alternative analyses were suggested to me by Zoltán Kövecses which I gratefully acknoweldge. The analyses outlined here follow from considerations put forward in Kövecses (2000).

2. Although this problem is exemplified by Chapter 8 here, it applies to all of the case studies.

3. In what follows we will make use of some of the aspects of thought experiments discussed in Cooper (1999).

4. There have been serious attempts to handle inconsistency in science which should be mentioned. In general see e.g. Rescher (1987), Rescher and Brandom (1980), Priest (1998) etc. For a recent overview see Meheus (ed.) (2002).

5. For a systematic attempt see Kertész (2004), Kertész and Rákosi (2004). The role of inconsistencies in syntactic argumentation is discussed very instructively in Moravcsik (1993) and (2004).

6. This attitude emphasizes the essential openness and pluralism of the cognitive science of science as well as the need for open-mindedness and tolerance.

References

Achinstein, P. 1983. *The Nature of Explanation.* Oxford: Oxford University Press.

Allwood, J. and Gärdenfors, P. (eds.). 1999. *Cognitive Semantics: Meaning and Cognition.* Amsterdam and Philadelphia: John Benjamins.

Almeder, R. 1990. "On Naturalizing Epistemology". *American Philosophical Quarterly* 27: 263–279.

Aoun, J. and Sportiche, D. 1983. "On the Formal Theory of Government". *The Linguistic Review* 2: 211–236.

Arbib, M. A. and Hesse, M. B. 1986. *The Construction of Reality.* Cambridge: Cambridge University Press.

Auroux, S. and Kouloughli, D. 1993. "Why is there no 'true' philosophy of linguistics?". In *Linguistics and Philosophy: The Controversial Interface,* R. Harré and R. Harris (eds.), (1993), 21–44. Oxford: Pergamon.

Baldauf, Ch. 1997. *Metapher und Kognition. Grundlagen einer neuen Theorie der Alltagsmetapher.* Frankfurt am Main, Bern, New York and Paris: Peter Lang.

Barnes, B., Bloor, D. and Henry, J. 1996. *Scientific Knowledge. A Sociological Analysis.* London: Athlone.

Bechtel, M. and Graham, G. (eds.). 1999. *A Companion to Cognitive Science.* Oxford: Blackwell.

Bibok, K. and Németh T., E. (eds.). 2001. *Pragmatics and the Flexibility of Word Meaning.* Amsterdam, Oxford, New York, Tokyo and Singapore: Elsevier.

Bierwisch, M. 1979. "Wörtliche Bedeutung: Eine pragmatische Gretchenfrage". In *Sprechakttheorie und Semantik,* G. Grewendorf (ed.), 119–148. Frankfurt am Main: Suhrkamp.

Bierwisch, M. 1980. "Semantic Structure and Illocutionary Force". In *Speech Act Theory and Pragmatics,* J. R. Searle, F. Kiefer, and M. Bierwisch (eds.), 1–36. Dordrecht: Reidel.

Bierwisch, M. 1981. *Die Integration autonomer Systeme. Überlegungen zur kognitiven Linguistik.* Diss. Berlin.

Bierwisch, M. 1983a. *Essays in the Psychology of Language.* Berlin: Zentralinstitut für Sprachwissenschaft.

Bierwisch, M. 1983b. "Semantische und konzeptuelle Repräsentation von lexikalischen Einheiten". In *Untersuchungen zur Semantik,* R. Růžička and W. Motsch (eds.), 61–100. Berlin: Akademie Verlag.

Bierwisch, M. 1987. "Linguistik als kognitive Wissenschaft. Erläuterungen zu einem Forschungsprogramm". *Zeitschrift für Germanistik* 8: 645–667.

Bierwisch, M. 1990. "Perspectives on Mind, Brain, and Language: Linguistics as a Cognitive Science or Touring the Chinese Room Again". In *Speech Acts, Meaning and Intentions:*

Critical Approaches to the Philosophy of J. R. Searle, A. Burckhardt (ed.), 391–428. Berlin and New York: de Gruyter.

Bierwisch, M. 1999a. "Das Organ des Denkens und die Grenzen des Ausdrückbaren". In *"Werkzeug Sprache". Sprachpolitik, Sprachfähigkeit, Sprache und Macht. 3. Symposion der deutschen Akademien der Wissenschaften*, Union der Deutschen Akademien der Wissenschaften zu Leipzig (ed.), 57–101. Hildesheim, Zürich and New York: Olms.

Bierwisch, M. 1999b. "Words as Programs of Mental Computation". In *Learning — Rule Extraction and Representation*, A. D. Friederici and R. Menzel (eds.), 3–35. Berlin and New York: de Gruyter.

Bierwisch, M. and Lang, E. 1989. "Somewhat Longer — Much Deeper — Further and Further: Epilogue to the Dimensional Adjective Project". In M. Bierwisch and E. Lang (eds.), (1989), 471–514.

Bierwisch, M. and Schreuder, R. 1992. "From Concepts to Lexical Items". *Cognition* 42: 23–60.

Bierwisch, M. and Lang, E. (eds.). 1989. *Dimensional Adjectives: Grammatical Structure and Conceptual Interpretation.* Berlin: Springer.

Bierwisch, M. and Bosch, P. (eds.). 1995. *Semantic and Conceptual Knowledge.* Stuttgart and Tübingen: IBM Deutschland.

Bird, A. 1998. *Philosophy of Science.* Montreal & Kingston, London, Buffalo: McGill-Queen's University Press.

Bishop, M.A. 1999. "Why Thought Experiments are not Arguments". *Philosophy of Science* 66: 534–541.

Bloor, D. 1976. *Knowledge and Social Imagery.* London: Routledge.

Bloor, D. 1992. "Ordinary Human Inference as Material for the Sociology of Knowledge". *Social Studies of Science* 21: 129–139.

Bloor, D. 1983. *Wittgenstein: A Social Theory of Knowledge.* London: Macmillan.

Bloor, D. 1997. *Wittgenstein, Rules and Institutions.* London: Routledge.

Boas, H. U. 1984. *Formal Versus Explanatory Generalizations in Generative Transformational Grammar: An Investigation into Generative Argumentation.* Tübingen: Niemeyer.

Borsboom, D., Mellenbergh, G.J and van Heerden, J. 2002. "Functional Thought Experiments". *Synthese* 130: 379–387.

Botha, R. P. 1981. *The Conduct of Linguistic Inquiry: A Systematic Introduction to the Methodology of Generative Grammar.* The Hague: Mouton.

Börkel, M. 1995. "Conceptual Interpretation of Systematic Polysemy". In M. Bierwisch and P. Bosch (eds.), (1995), 69–78.

Brannigan, A. 1981. *The Social Basis of Scientific Discoveries.* Cambridge: Cambridge University Press.

Brown, J. R. 1991. *The Laboratory of the Mind.* London: Routledge.

Brown, J. R. 2000. *Who Should Rule? A Guide to the Epistemology and Politics of the Science Wars.* New York and Berlin: Springer.

Brown, J. R. 2001. "Social Factors in Science". In W. H. Newton-Smith (ed.), (2001), 442–450.

Brown, J. R. (ed.). 1984. *Scientific Rationality: The Sociological Turn.* Dordrecht: Reidel.

Carnap, R. and Stegmüller, W. 1959. *Induktive Logik und Wahrscheinlichkeit.* Wien: Springer.

Carr, D. 1990. *Linguistic Realities.* Cambridge: Cambridge University Press.

Carstensen, C.-U. 1995. "Semantic and Conceptual Aspects of Local Expressions: Critical Remarks on the 'State of the Art'". In M. Bierwisch and P. Bosch (eds.), (1995), 117–126.
Chomsky, N. 1957. *Syntactic Structures.* The Hague: Mouton.
Chomsky, N. 1964. *Current Issues in Linguistic Theory.* The Hague: Mouton.
Chomsky, N. 1965. *Aspects of the Theory of Syntax.* Cambridge, MA: The MIT Press.
Chomsky, N. 1973. "Conditions on Transformations". In *A Festschrift for Morris Halle*, S. R. Anderson and P. Kiparsky (eds.), 232–286. New York: Holt, Rinehart and Winston.
Chomsky, N. 1979. *Language and Responsibility.* New York: Pantheon.
Chomsky, N. 1980. "On Binding". *Linguistic Inquiry* 11: 1–46.
Chomsky, N. 1981. *Lectures on Government and Binding.* Dordrecht: Foris.
Chomsky, N. 1986. *Knowledge of Language: Its Nature, Origin and Use.* New York: Praeger.
Chomsky, N. 1987. *Barriers.* Cambridge, MA: The MIT Press.
Chomsky, N. 1991. "Linguistics and Adjacent Fields: A Personal View". In A. Kasher (ed.), (1991), 3–25.
Churchland, P. M. 1989. *A Neurocomputational Perspective: The Nature of Mind and the Structure of Science.* Cambridge, MA: The MIT Press.
Churchland, P. 1986. *Neurophilosophy: Toward a Unified Science of the Mind/Brain.* Cambridge, MA: The MIT Press.
Cohen, D. (ed.). 1974. *Explaining Linguistic Phenomena.* New York: Wiley.
Cohen, D. and Wirth, J. (eds.). 1975. *Testing Linguistic Hypotheses.* New York: Wiley.
Collins, H. M. 1985. *Changing Order: Replication and Induction in Scientific Practice.* Beverly Hills, CA: Sage.
Collins, H. M. 1990. *Artificial Experts: Social Knowledge and Intelligent Machines.* Cambridge, MA: The MIT Press.
Cooper, R. 1999. "From Thought Experiments to Real Experiments: Methodology in the Philosophy of Mind". *Studies in the History and Philosophy of Science* 30: 263–271.
Dancy, J. and Sosa, E. (eds.). 1997. *A Companion to Epistemology.* Oxford: Blackwell.
Davidson, D. and Hintikka, J. (eds.). 1969. *Words and Objections: Essays on the Work of W. V. Quine.* Dordrecht: Reidel.
d'Avis, W. 1998. "Theoretische Lücken der Cognitive Science". *Journal for General Philosophy of Science* 29: 37–57.
Dirven, R. 2000. "Cognitive Linguistics". *LAUD* No. 515: 1–12. [To appear in: *The Linguistics Encyclopedia*, 2nd ed., K. Malmkjaer (ed.), London: Routledge.]
Downes, S. 1993. "Socializing Naturalized Philosophy of Science". *Philosophy of Science* 60: 452–468.
Drewer, P. 2003. *Die kognitive Metapher als Werkzeug des Denkens. Zur Rolle der Analogie bei der Gewinnung und Vermittlung wissenschaftlicher Erkenntnisse.* Tübingen: Narr.
Dupré, J. 1982. The Disunity of Science. *Mind* 93: 321–346.
Dupré, J. 1993. *The Disorder of Things: Metaphysical Foundations of the Disunity of Science.* Cambridge, MA: Harvard University Press.
Durkheim, E. 1936. *The Rules of Sociological Method.* New York: The Free Press.
Elfers, E. 2003. "Generative Grammar: A Research Programme?" In *Später Mittag. Vermischte Anmerkungen zur Metahistoriographie. Festgabe für Peter Schmitter zum 60. Geburtstag*, K. D. Dutz (ed.),65–80. Münster: Nodus.

Eemeren, F. H. van and Grootendorst, R. 1992. *Argumentation, Communication and Fallacies.* Hillsdale: Erlbaum.

Fauconnier, G. 1985. *Mental Spaces.* Cambridge, MA: The MIT Press.

Fehér, M. 1988. "Epistemology Naturalized vs. Epistemology Socialized". In *Scientific Knowledge Socialized. Selected Proceedings of the 5th Joint International Conference on the History and Philosophy of Science*, I. Hronszky, M. Fehér and B. Dajka (eds.), 75–98. Dordrecht: Kluwer.

Fehér, M. 1990. "The Essential Tension: On the Role of Inconsistencies in Science". *Studies in Soviet Thought* 39: 231–239.

Fillmore, Ch. 1985. "Frames and Semantics of Understanding". *Quaderni di Semantica* 62: 222–253.

Finke, P. 1979. *Grundlagen einer linguistischen Theorie. Empirie und Begründung in der Sprachwissenschaft.* Braunschweig: Vieweg.

Finke, P. 1982. *Konstruktiver Funktionalismus. Die wissenschaftstheoretische Basis einer empirischen Theorie der Literatur.* Braunschweig: Vieweg.

Floridi, L. 1993. "The Problem of the Justification of a Theory of Knowledge (1): Some Historical Metamorphoses". *Journal for General Philosophy of Science* 24: 205–233.

Floridi, L. 1994. "The Problem of the Justification of a Theory of Knowledge (2): Morphology and Diagnosis". *Journal for General Philosophy of Science* 25: 17–49.

Floridi, L. 1996. *Scepticism and the Foundation of Epistemology. A Study in the Metalogical Fallacies.* Köln, Leiden and New York: Brill.

Fodor, J. A. 1983. *The Modularity of Mind.* Cambridge, MA: The MIT Press.

Forrai, G, 1987. "The Role of Metaphor in the Birth of Generative Grammar". *Doxa* 9: 49–56. [Budapest].

Fuller, S. 1993a. "David Bloor's 'Knowledge and Social Imagery'". *Philosophy of Science* 60: 158–170.

Fuller, S. 1993b. *Philosophy, Rhetoric and the End of Knowledge: The Coming of Science and Technology Studies.* Madison: The University of Wisconsin Press.

Fuller, S., De Mey, M., Shinn, T. and Woolgar, S. (eds.). 1989. *The Cognitive Turn: Sociological and Psychological Perspectives on Science.* Dordrecht: Kluwer.

Gardner, H. 1985. *The Mind's New Science. A History of the Cognitive Revolution.* New York: Basic Books.

Geeraerts, D. 1999. "Idealist and Empiricist Tendencies in Cognitive Semantics". In Th. Janssen and G. Redeker (eds.), 163–194.

Gendler, T.S. 1998. "Galileo and the Indispensability of Scientific Thought Experiment". *British Journal for the Philosophy of Science* 49: 397–424.

Gibbs, R. W. 1996. "What's Cognitive About Cognitive Linguistics?". In *Linguistics in the Redwoods: The Expansion of a New Paradigm in Linguistics*, E. H. Casad (ed.), 27–53. Berlin and New York: de Gruyter.

Giere, R. 1988. *Explaining Science: A Cognitive Approach.* Chicago: University of Chicago Press.

Giere, R. 2001a. "Cognitive Approaches to Science". In W. H. Newton-Smith (ed.), (2001), 41–43.

Giere, R. 2001b. "Naturalism". In W. H. Newton-Smith (ed.), 308–310.

Goldberg, A. E. 1995. *Construction Grammar Approach to Argument Structure.* Chicago: University of Chicago Press.
Goldman, A. I. 1986. *Epistemology and Cognition.* Chicago: Harvard University Press.
Grewendorf, G., Hamm, F. and Sternefeld, W. 1987. *Sprachliches Wissen. Eine Einführung in moderne Theorien der grammatischen Beschreibung.* Frankfurt am Main: Suhrkamp.
Gutfleisch-Rieck, I., Klein, W., Speck, A. and Spranz-Fogasy, Th. 1989. *Transkriptionsvereinbarungen für den Sonderforschungsbereich 245 "Sprechen und Sprachverstehen im sozialen Kontext".* Heidelberg: Universität Heidelberg.
Haack, S. 1993. "The Two Faces of Quine's Naturalism". *Synthese* 94: 335–356.
Hamblin, C. I. 1970. *Fallacies.* London: Methuen.
Hanna, J. F. 1968. "An Explication of 'Explication'". *Philosophy of Science* 35: 28–44.
Harder, P. 1999. "Partial Autonomy. Ontology and Methodology in Cognitive Linguistics". In Th. Janssen and G. Redeker (eds.), 195–222.
Harré, R. and Harris, R. 1993. "Preface". In *Linguistics and Philosophy: The Controversial Interface,* R. Harré and R. Harris (eds.), ix-xi. Oxford: Pergamon.
Haser, V. 2003. "Theo Janssen and Gisela Redeker (eds.) 'Cognitive Linguistics: Foundations, Scope and Methodology'. Review". *Cognitive Linguistics* 14: 71–86.
Helman, D. H. (ed.). 1988. *Analogical Reasoning.* Dordrecht: Kluwer.
Hempel, C. G. 1965. *Aspects of Scientific Explanation.* New York: Free Press.
Hesse, M. 1966. *Models and Analogies in Science.* Notre Dame: Notre Dame University Press.
Hesse, M. 1988. "Theory, Family Resemblances and Analogy". In D. H. Helman (ed.), (1998), 317–340.
Hesse, M. 1980. *Revolutions and Reconstructions in the Philosophy of Science.* Brighton: The Harvester Press.
Huby, P. and Neal, G. (eds.). 1989. *The Criterion of Truth.* Liverpool: Liverpool University Press.
Hume, D. 1739. A Treatise of Human Nature. Books 1 and 2 (repr. London: Fontana Books).
Itkonen, E. 1978. *Grammatical Theory and Metascience.* Amsterdam and Philadelphia: John Benjamins.
Itkonen, E. 1983. *Causality in Linguistic Theory.* London: Croom Helm.
Jackendoff, R. 1983. *Semantics and Cognition.* Cambridge, MA: The MIT Press.
Jackendoff, R. 1990. *Semantic Structures.* Cambridge, MA: The MIT Press.
Jackendoff, R. and Aaron, D. 1991. "'More than Cool Reason: A Field Guide to Poetic Metaphor, by George Lakoff and Mark Turner.' Review Article". *Language* 67: 320–338.
Janssen, Th. and Redeker, G. (eds.). 1999. *Cognitive Linguistics: Foundations, Scope, and Methodology.* Berlin and New York: de Gruyter.
Jäkel, O. 1996. "Metaphorical Scenarios of Science". In *The Construal of Space in Language and Thought,* M. Pütz and R. Dirven (eds.), 649–678. Berlin and New York: de Gruyter.
Jäkel, O. 1997. *Metaphern in abstrakten Diskurs-Domänen.* Frankfurt am Main, Bern, New York and Paris: Lang.
Johnson, M. 1987. *The Body in the Mind. The Bodily Basis of Meaning, Imagination, and Reason.* Chicago and London: The University of Chicago Press.

Johnson-Laird, P. N. 1980. "Mental Models in Cognitive Science". *Cognitive Science* 4: 71–115.

Johnson-Laird, P. N. 1983. *Mental Models.* Cambridge: Cambridge University Press.

Junker, M-O. 1992. "Metaphors We Live By: The Terminology of Linguistic Theory". *Natural Language and Linguistic Theory* 10: 142–145.

Kasher, A. (ed.). 1991. *The Chomskyan Turn.* Oxford: Blackwell.

Kertész, A. 1988. "Zur Bewertung der pragmatischen Erklärungsmodelle". *Zeitschrift für allgemeine Wissenschaftstheorie* 19: 239–251.

Kertész, A. 1991. *Die Modularität der Wissenschaft. Konzeptuelle und soziale Prinzipien linguistischer Erkenntnis.* Braunschweig: Vieweg.

Kertész, A. 1993. *Artificial Intelligence and the Sociology of Knowledge: Prolegomena to an Integrated Philosophy of Science.* Frankfurt am Main, Bern, New York and Paris: Lang.

Kertész, A. 2001a. "Pragmatics and the Flexibility of Theoretical Terms in Linguistics: Two Case Studies". In K. Bibok and E. Németh T. (eds.), (2001), 117–150.

Kertész, A. 2001b. "Introduction". In A. Kertész (ed.), (2001), 9–15.

Kertész, A. 2001c. "Metascience and the Metaphorical Structure of Scientific Discourse". In A. Kertész (ed.), (2001), 135–158.

Kertész, A. 2002a. "On the Contribution of Metascience to Cognitive Linguistics: A Case Study". *Linguistische Berichte* 190: 207–228.

Kertész, A. 2002b. "On the De-Naturalization of Epistemology". *Journal for General Philosophy of Science* 33: 269–288.

Kertész, A. 2004. *Philosophie der Linguistik. Studien zur naturalisierten Wissenschaftstheorie.* Tübingen: Narr.

Kertész, A. and Rákosi, Cs. 2004. "Plausible Reasoning in a Theory of German Affricates: A Case Study". *Sprachtheorie und germanistische Linguistik* 14 (to appear)

Kertész, A. (ed.). 2001. *Approaches to the Pragmatics of Scientific Discourse.* Frankfurt am Main, Bern, Oxford, New York and Wien: Lang.

Kiefer, F. 1995. "Cognitive Linguistics: A New Paradigm?". In *Linguistics in the Morning Calm 3,* I. H. Lee (ed.), 93–110. Seoul: Hanshin.

Kitcher, P. 1992. "The Naturalists Return". *The Philosophical Review* 101: 53–114.

Kim, J. 1988. "What is 'Naturalized Epistemology'?". In *Philosophical Perspectives, 2: Epistemology.* J. Tomberlin (ed.), 381–406. Atascaredo, CA: Ridgeview.

Klima, E. S. 1964. "Negation in English". In *The Structure of Language. Readings in the Philosophy of Language,* J. A. Fodor and J. J. Katz (eds.), 246–323. Englewood Cliffs, NJ: Prentice Hall.

Knorr-Cetina, K. 1981. *The Manufacture of Knowledge: An Essay on the Constructivist and Contextual Nature of Knowledge.* Oxford: Pergamon.

Koppelberg, D. 1996. "Was macht eine Erkenntnistheorie naturalistisch?" *Journal for General Philosophy of Science* 27: 71–90.

Kornblith, H. (ed.). 1985. *Naturalized Epistemology.* Dordrecht: Reidel.

Kornblith, H. 1993. "Epistemic Normativity". *Synthese* 94: 357–376.

Kövecses, Z. 1986. *Metaphors of Anger, Pride, and Love. A Lexical Approach to the Study of Concepts.* Amsterdam and Philadelphia: John Benjamins.

Kövecses, Z. 1990. *Emotion Concepts.* Berlin, Heidelberg and New York: Springer.

Kövecses, Z. 2000. *Metaphor and Emotion.* Cambridge: Cambridge University Press.

Kuhn, T. S. 1970. *The Structure of Scientific Revolutions.* Chicago: Chicago University Press.

Kuhn, T. S. 1981. "A Function for Thought Experiments". In *Scientific Revolutions.* I. Hacking (ed.), 6–27. Oxford: Oxford University Press.

Lakoff, G. 1987. *Women, Fire, and Dangerous Things: What Categories Reveal about the Mind.* Chicago: University of Chicago Press.

Lakoff, G. 1990. "The Invariance Hypothesis: Is Abstract Reason Based on Image Schemas?". *Cognitive Linguistics* 1: 39–74.

Lakoff, G. 1993. "The Contemporary Theory of Metaphor". In *Metaphor and Thought,* A. Ortony (ed.), 202–252. Cambridge: Cambridge University Press.

Lakoff, G. and Johnson, M. 1980a. *Metaphors We Live By.* Chicago and London: University of Chicago Press.

Lakoff, G. and Johnson, M. 1980b. "The Metaphorical Structure of the Human Conceptual System". *Cognitive Science* 4: 195–208.

Lakoff, G. and Johnson, M. 1999. *Philosophy in the Flesh. The Embodied Mind and Its Challange to Western Thought.* New York: Basic Books.

Lakoff, G. and Turner, M. 1989. *More Than Cool Reason: A Field Guide to Poetic Metaphor.* Chicago and London: The University of Chicago Press.

Lakoff, G. and Núñez, R. 2002. *Where Mathematics Comes From: How the Embodied Mind Brings Mathematics Into Being.* New York: Basic Books.

Lang, E. 1989. "The Semantics of Dimensional Designation". In M. Bierwisch and E. Lang (eds.), (1989), 263–458.

Lang, E. 1991. "Linguistische vs. konzeptuelle Aspekte der LILOG-Ontologie". In *Ontologie und Axiomatik der Wissensbasis von LILOG,* G. Klose, E. Lang, and Th. Pirlein (eds.), 23–46. Berlin, Heidelberg, New York, Paris etc.: Springer.

Lang, E. 1994. "Semantische vs. konzeptuelle Struktur: Unterscheidung und Überschneidung". In M. Schwarz (ed.), (1994), 25–40.

Lang. E. and Carstensen, U. 1990. *OSKAR B A PROLOG Program for Modelling Dimensional Designation and Positional Variation of Objects in Space.* Stuttgart: IBM Deutschland.

Lang, E., Carstensen, K.-U. and Simmons, G. 1991. *Modelling Spatial Knowledge on a Linguistic Basis: Theory – Prototype – Integration.* Berlin, Heidelberg and New York: Springer.

Langacker, R. W. 1969. "On Pronominalization and the Chain of Command". In *Modern Studies in English: Readings in Transformational Grammar,* D. A. Reibel and S. A. Schane (eds.), 160–186. Englewood Cliffs, NJ: Prentice Hall.

Langacker, R. W. 1987. *Foundations of Cognitive Grammar. Vol. 1.* Stanford, CA: Stanford University Press.

Langacker, R. W. 1991a. *Foundations of Cognitive Grammar. Vol. 2.* Stanford, CA: Stanford University Press.

Langacker, R. W. 1991b. *Concept, Image, Symbol.* Berlin and New York: de Gruyter.

Langacker, R. W. 1999a. *Grammar and Conceptualization.* Berlin and New York: de Gruyter.

Langacker, R. W. 1999b. "Assessing the Cognitive Linguistic Enterprise". In Th. Janssen and G. Redeker (eds.), (1999), 13–60.

Langley, P., Simon, H., Bradshaw, G. L. and Zytkow, J. M. (eds.). 1987. *Scientific Discovery: Computational Explorations of the Creative Process.* Cambridge, MA: The MIT Press.

Lasnik, H. 1976. "Remarks on Coreference". *Linguistic Analysis* 2: 1–22.
Latour, B. 1987. *Science in Action.* Cambridge, MA: Harvard University Press.
Latour, B. 1988. *The Pasteurization in France.* Cambridge, MA: Harvard University Press.
Latour, B. and Woolgar, S. 1979. *Laboratory Life: The Social Construction of Scientific Facts.* Beverly Hills: Sage.
Laudan, L. 1984. "The Pseudo-Science of Science?". In J. R. Brown (ed.), (1984), 41–74.
Liebert, W.-A. 1995. "The Lexicon of Metaphor Models as a Mental Tool for Analogical Problem-Solving in Science". In *Current Approaches to the Lexicon*, R. Dirven and J. Vanparys (eds.), 433–448. Frankfurt am Main, Bern, New York and Paris: Lang.
Liebert, W.-A. 1996a. "Hypertextdesign in der kognitiven Lexikographie". In *Wörterbücher in der Diskussion II*, H.-E. Wiegand (ed.), 103–139. Tübingen: Niemeyer.
Liebert, W.-A. 1996b. "Die transdiskursive Vorstellungswelt zum AIDS-Virus". In *Fachliche Textsorten*, H. Kalverkämper and K. D. Baumann (eds.), 789–811. Tübingen: Niemeyer.
Liebert, W.-A. 1997a. "Stop Making Sense! Metaphor and Perspective in Creative Thinking Sessions of Scientists and Scientific Radio Broadcasts". In *Discourse and Perspective in Cognitive Linguistics*, W. A. Liebert, G. Redeker and L. Waugh (eds.), 149–183. Amsterdam and Philadelphia: John Benjamins.
Liebert, W.-A. 1997b. "Interaktion und Kognition. Die Herausbildung metaphorischer Denkmodelle in Gesprächen zwischen Wissenschaftlern und Wissenschaftsjournalisten". In *Metaphern, Medien, Wissenschaft. Zur Vermittlung der AIDS-Forschung in Presse und Rundfunk*, B. U. Biere and W.-A. Liebert (eds.), 180–209. Opladen: Westdeutscher Verlag.
Long, A. A. 1989. "Ptolemy on the Criterion: an Epistemology for the Practicing Scientist". In P. Huby and G. Neal (eds.), (1989), 151–172.
Maffie, J. 1990. "Recent Work on Naturalized Epistemology". *American Philosophical Quarterly* 27: 281–293.
Maffie, J. 1995. "Naturalism, Scepticism and the Independence of Epistemology". *Erkenntnis* 43: 1–27.
Mannheim, K. 1936. *Ideology and Utopia.* London: Routledge and Kegan Paul.
May, R. 1985. *Logical Form: Its Structure and Derivation.*Cambridge, MA: The MIT Press.
McGlone, M. S. 2001. "Concepts as Metaphors". In *Understanding Figurative Language. From Metaphors to Idioms*, S. Glucksberg, 97–107. Oxford: Oxford University Press.
Meheus, J. (ed.). 2002. *Inconsistency in Science.* Dordecht: Kluwer.
Moravcsik, E. 1993. Why Is Syntax complicated? In *Principles and Prediction: The Analysis of Natural Language*, M. Eid and G. Iverson (eds.), 73–92, Amsterdam and Phildadelphia: John Benjamins.
Moravcsik, E. 2004. *Syntax. An Introduction.* Amsterdam and Philadelphia: John Benjamins (to appear).
Müller, R.-A. 1991. *Der (un)teilbare Geist. Modularismus und Holismus in der Kognitionsforschung.* Berlin and New York: de Gruyter.
Murphy, G. L. 1996. "On Metaphoric Representation". *Cognition* 60: 173–186.
Murphy, G. L. 1997. "Reasons to Doubt the Present Evidence for Metaphoric Representation". *Cognition* 62: 99–108.
Neisser, U. 2001. "George Lakoff and Mark Johnson, 'Philosophy in the Flesh'. Review".

Language 77: 166–169.

Nersessian, N. 1992. "In the Theoretician's Laboratory: Thought Experiments as Mental Modeling". In *Proceedings of the 1992 Biennial Meeting of the Philosophy of Science Association. Vol. 2*, D. Hull, M. Forbes and K. Okruhlik (eds.), 129–301. Michigan: Philosophy of Science Association.

Newmeyer, F. J. 1986. *Linguistic Theory in America.* New York: Academy Press.

Newmeyer, F. J. 1991. "Rules and Principles in the Historical Development of Generative Syntax". In A. Kasher (ed.), (1991), 200–230.

Newton-Smith, W. H. 2001a. "Introduction". In W. H. Newton-Smith (ed.), (2001), 1–8.

Newton-Smith, W. H. 2001b. "Explanation". In W. H. Newton-Smith (ed.), (2001), 127–133.

Newton-Smith, W. H. (ed.). 2001. *A Companion to the Philosophy of Science.* Oxford: Blackwell.

Nickles, T. 1989. "Integrating the Science Studies Disciplines'". In S. Fuller et al. (eds.), (1989), 225–256.

Nola, R. 1992. "Ordinary Human Inference as the Refutation of the Strong Programme". *Social Studies of Science* 21: 107–129.

Norton, J. 1991. "Thought Experiments in Einstein's Work". In *Thought Experiments in Science and Philosophy*, T. Horowitz and G. Massey (eds.), 129–148. Savage: Rowman and Littlefield.

Norton, J.D. 1996. "Are Thought Experiments Just What You Thought?". *Canadian Journal of Philosophy* 26: 333–366.

Oesterreicher, W. 1979. *Sprachtheorie und Theorie der Sprachwissenschaft.* Heidelberg: Winter.

Peijnenburg, J. and Atkinson, D. 2003. "When are Thought Experiments Poor Ones?". *Journal for General Philosophy of Science* 34: 305–322.

Percival, P. 2001. "Theoretical Terms: Meaning and Reference". In W. H. Newton-Smith (ed.) (2001), 495–514.

Peregrin, J. 1998. "Linguistics and Philosophy". *Theoretical Linguistics* 24: 245–264.

Perry, Th.A. (ed.). 1980. *Evidence and Argumentation in Linguistics.* Berlin and New York: de Gruyter.

Pethő, G. 2001a. "Konzeptuelle Fokussierung. Bemerkungen zur Behandlung der Polysemie in der Zwei-Ebenen Semantik". In *Wort und (Kon)text*, P. Kocsány and A. Molnár (eds.), 19–57. Frankfurt am Main, Bern, Bruxelles, Oxford and New York: Lang.

Pethő, G. 2001b. "What is Polysemy? A Survey of Current Research and Results". In K. Bibok and E. Németh T. (eds.), (2001), 175–224.

Pickering, A. 1984. *Constructing Quarks.* Chicago: University of Chicago Press.

Polya, G. 1948. *How to Solve It?* Princeton: Princeton University Press.

Polya, G. 1954. *Mathematics and Plausible Reasoning. Vol. II. Patterns of Plausible Inference.* Oxford: Oxford University Press.

Priest, G. 1998. "What is so bad about contradictions?". *The Journal of Philosophy* XCV: 410–426.

Putnam, H. 1962. "What Theories Are Not". In *Logic, Methodology and Philosophy of Science*, E. Nagel, P. Suppes and A. Tarski (eds.), (1962). Stanford, CA.

Quine, W. V. O. 1960. *Word and Object.* Cambridge, MA: The MIT Press.

Quine, W. V. O. 1969a. "Epistemology Naturalized". In *Ontological Relativity and Other Essays*, W. V. O. Quine (ed.), 69–90. New York: Columbia University Press.

Quine, W. V. O. 1969b. "Reply to Smart". In D. Davidson and J. Hintikka (eds.), (1969), 292–294.

Quine, W. V. O. 1981. *Theories and Things*. Cambridge, MA: Harvard University Press.

Quine, W. V. O. 1986. "Reply to Morrison White". In *The Philosophy of W. V. Quine*, L. Hahn and P. Shilpp (eds.), 663–65.

Redeker, G. and Janssen, Th. 1999. "Introduction". In Th. Janssen and G. Redeker (eds.), 1–12.

Reinhart, T. 1974. "Syntax and Coreference". In *NELS* 3: 92–105.

Reinhart, T. 1983a. "Coreference and Bound Anaphora: A Restatement of the Anaphora Questions". *Linguistics and Philosophy* 6: 47–88.

Reinhart, T. 1983b. *Anaphora and Semantic Interpretation*. London: Croom Helm.

Reis, M. 1987. "Die Stellung der Verbargumente im Deutschen: Stilübungen zum Grammatik-Pragmatik-Verhältnis". In *Sprache und Pragmatik. Lunder Symposium 1986*, I. Rosengren (ed.), (1987), 139–178. Stockholm: Almsell and Wiksell.

Rescher, N. 1976. *Plausible Reasoning*. Assen and Amsterdam: Van Gorcum.

Rescher, N. 1987. "How Serious a Fallacy is Inconsistency?". *Argumentation* 1: 303–316.

Rescher, N. and Brandom, R. 1980. *The Logic of Inconsistency*. Oxford: Blackwell.

Riley, K. 1987. "The Metalanguage of Transformational Syntax: Relations between Jargon and Theory". *Semiotica* 67: 173–194.

Richards, I. A. 1971 [1936)]. *The Philosophy of Rhetoric*. New York: Oxford University Press.

Rouse, J. 1998. "New Philosophies of Science in North America — Twenty Years Later". *Journal for General Philosophy of Science* 29: 71–122.

Ricoeur, P. 1986 [1978]. *The Rule of Metaphor: Multi-disciplinary Studies of the Creation of Meaning in Language*. London: Routledge and Kegan Paul.

Rudzka-Ostyn, B. (ed.). 1988. *Topics in Cognitive Linguistics*. Amsterdam and Philadelphia: John Benjamins.

Saferstein, B. 1999. "Ethnomethodology". In M. Bechtel and G. Graham (eds.), (1999), 477–487.

Salmon, W. 1997. "Explanation". In J. Dancy and E. Sosa (eds.), (1997), 129–132.

Sas, P. J. 1999. "Plugging the Leaks in Neurath's Ship. A Defence of Naturalistic Epistemology". *Journal for General Philosophy of Science* 30: 131–153.

Schwarz, M. 1992. *Einführung in die Kognitive Linguistik*. Tübingen: Francke.

Schwarz, M. 1997. "Kognitive Linguistik? Eine Straße in den Geist!". In *Metalinguistik im Wandel. Die 'kognitive Wende' in Wissenschaftstheorie und Linguistik*, A. Kertész (ed.), 19–30. Frankfurt am Main, Bern, New York and Paris: Lang.

Schwarz, M. 2002. "Konzeptuelle Ansätze II: Einebenen-Ansatz vs. Mehrebenen-Ansatz". In *Lexikologie. Lexicology. Ein internationales Handbuch zur Natur und Struktur von Wörtern und Wortschätzen. An International Handbook on the Nature and Structure of Words and Vocabularies*, D. A. Cruse, F. Hundsnurscher, M. Job, R. P. Lutzeier (eds.), 277–384. Berlin and New York: de Gruyter.

Schwarz, M. (ed.). 1994. *Kognitive Semantik/Cognitive Semantics. Ergebnisse, Probleme, Perspektiven*. Tübingen: Narr.

Scriven, M. 1962. "Explanations, Predictions, Laws". In *Minnesota Studies in the Philosophy of Science. Vol. 3.* 172–230. Minneapolis: University of Minnesota Press.

Shapin, S. and Schaffler, S. 1985. *Leviathan and the Air Pump.* Princeton: Princeton University Press.

Shatz, D. 1993. "Scepticism and Naturalized Epistemology". In *Naturalism: A Critical Reappraisal,* S. J. Wagner and R. Warner (eds.), (1993), 117–145. Notre Dame, IN: Notre Dame University Press.

Slezak, P. 1989. "Scientific Discovery by Computer as Empirical Refutation of the Strong Programme". *Social Studies of Science* 19: 563–600.

Sneed, J. D. 1971. *The Logical Structure of Mathematical Physics.* Dordrecht: Reidel.

Sokal, A. 1996a. "Transgressing the Boundaries: Toward a Transformative Hermeneutics of Quantum Gravity". *Social Text* 46–47: 217–253.

Sokal, A. 1996b. "A Physicist Experiments With Cultural Studies". *Lingua Franca,* May-June: 62–64.

Sokal, A. and Bricmont, J. 1998. *Intellectual Impostures.* London: Profile Books.

Sorensen, R. 1992. *Thought Experiments.* Oxford: Oxford U.P.

Sperber, D. and Wilson, D. 1986. *Relevance: Communication and Cognition.* Oxford: Blackwell.

Stechow, A. von and Sternefeld, W. 1988. *Bausteine syntaktischen Wissens: Ein Lehrbuch der generativen Grammatik.* Opladen: Westdeutscher Verlag.

Stegmüller, W. 1970. *Theorie und Erfahrung.* Berlin, Heidelberg and New York: Springer.

Stegmüller, W. 1978. *Personelle und statistische Wahrscheinlichkeit.* Berlin, Heidelberg and New York: Springer.

Stegmüller, W. 1983. *Erklärung — Begründung — Kausalität.* Berlin, Heidelberg and New York: Springer.

Stich, P. 1990. *The Fragmentation of Reason.*Cambridge, MA: The MIT Press.

Strawson, P. F. 1985. *Scepticism and Naturalism.* New York: Columbia University Press.

Striker, G. 1990. "The Problem of the Criterion". In *Epistemology,* S. Everson (ed.), 143–160. Cambridge, MA: Cambridge University Press.

Stump, D. 1993. "Naturalized Philosophy of Science with a Plurality of Methods". *Philosophy of Science* 59: 457–460.

Suppe, P. 1977. "The Search for Philosophic Understanding of Scientific Theories". In *The Structure of Scientific Theories,* P. Suppe (ed.), 1–241. Urbana, IL: The University of Illinois Press.

Sweetser, E. E. 1990. *From Etymology to Pragmatics: Metaphorical and Cultural Aspects of Semantic Structure.* Cambridge: Cambridge University Press.

Talmy, L. 1983. "How Language Structures Space". In *Spatial Orientation: Theory, Research, and Application,* H. L. Pick and L. P. Acredolo (eds.), 225–282. New York etc.: Plenum.

Talmy, L. 1988. "Force Dynamics in Language and Cognition". *Cognitive Science* 2: 49–100.

Taylor, J. R. 1994. "The Two-Level Approach to Polysemy". *Linguistische Berichte* 149: 3–26.

Taylor, J. R. 1995a. "Approaches to Word Meaning: The Network Model (Langacker) and the Two-Level Model (Bierwisch) in Comparison". In *Current Approaches to the Lexicon,* R. Dirven and J. Vanparys (eds.), 3–26. Frankfurt am Main, Bern, New York and Paris: Lang.

Taylor, J. R. 1995b. *Linguistic Categorization: Prototypes in Linguistic Theory.* Oxford: Oxford University Press.

Tomasello, M. 1999. "Cognitive Linguistics". In M. Bechtel and G. Graham (eds.), (1999), 477–487.

Thagard, P. 1988. *Computational Philosophy of Science.* Cambridge, MA: The MIT Press.

Thagard, P. 1992. *Conceptual Revolutions.* Princeton: Princeton University Press.

Tsiapera, M. 1990. "Organic Metaphor in Early 19th Century Linguistics". In *Historiography of Linguistics*, H.-J. Niederehe and E. F. K. Koerner (eds.), 577–587. Amsterdam and Philadelphia: John Benjamins.

Walton, D. N. 1987. *Informal Fallacies.* Amsterdam and Philadelphia: John Benjamins.

Walton, D. N. 1997. "Informal Fallacies". In J. Dancy and E. Sosa (eds.), (1997), 212–216.

Werner, H. 1985. "Die Semantik der Täuschungen". *Linguistische Berichte* 100: 477–510.

Woods, J. and Walton, D. N. 1992. *Fallacies: Selected Papers 1972–1982.* Dordrecht: Foris.

Wexler, K., Culicover, P. and Hamburger, H. 1974. Learning-Theoretic Foundations of Linguistic-Universals. *Theoretical Linguistics* 2: 215–253.

Wilkes, K. 1998. *Real People: Personal Identity Without Thought Experiments.* Oxford: Clarendon Press.

Williams, E. 1987. "Implicit Arguments, the Binding Theory and Control". *Natural Language and Linguistic Theory* 5: 151–180.

Wirth, J. (ed.). 1976. *Assessing Linguistic Arguments.* New York: Wiley.

Wunderlich, D. (ed.). 1976. *Wissenschaftstheorie der Linguistik.* Kronberg: Scriptor.

Zoglauer, Th. 1993. *Das Problem der theoretischen Terme.* Braunschweig: Vieweg.

Znaniecki, F. 1965. *The Social Role of the Man of Knowledge.* New York: Octagon Books.

Appendix

The metascientific extension of cognitive semantics (MECS): One of the tasks of cognitive semantics is to contribute to the solution of problems tackled by the naturalized philosophy of science in general and the cognitive science of science in particular.

The generalized modularity hypothesis (MH′): Human cognitive behaviour is organized in a modular way.

The metascientific extension of the generalized modularity hypothesis (MMH′): 'Scientific knowledge' is organized in a modular way along the lines of the generalized modularity hypothesis (MH′).

The generalized holistic hypothesis (HH): Human cognitive behaviour constitutes a unified system.

The main hypothesis of the cognitive theory of metaphor (HH′): Human cognitive behaviour is structured metaphorically.

The metascientific extension of the main hypothesis of the cognitive theory of metaphor (MHH′): 'Scientific knowledge' is structured by metaphorical concepts along the lines of the main hypothesis of the cognitive theory of metaphor (HH′).

Author index

Subject index

In the series *Converging Evidence in Language and Communication Research* the following titles have been published thus far or are scheduled for publication:

1 **HORIE, Kaoru (ed.):** Complementation. Cognitive and functional perspectives. 2000. vi, 242 pp.

2 **ALBERTAZZI, Liliana (ed.):** Meaning and Cognition. A multidisciplinary approach. 2000. vi, 270 pp.

3 **LOUWERSE, Max and Willie van PEER (eds.):** Thematics. Interdisciplinary Studies. 2002. x, 448 pp.

4 **KERTÉSZ, András:** Cognitive Semantics and Scientific Knowledge. Case studies in the cognitive science of science. 2004. xii, 259 pp.